Common Core Subject Test
Mathematics Grade 8

Student Practice Workbook

+ Two Full-Length Common Core Math Tests

Math Notion

www.MathNotion.com

Common Core Subject Test Mathematics Grade 8

Common Core Subject Test Mathematics Grade 8

Common Core Subject Test Mathematics Grade 8

Published in the United State of America By

The Math Notion

Web: WWW.MathNotion.com

Email: info@Mathnotion.com

Copyright © 2021 by the Math Notion. All rights reserved. No part of this publication may be reproduced, stored in a retrieval system, or transmitted in any form or by any means, electronic, mechanical, photocopying, recording, scanning, or otherwise, except as permitted under Section 107 or 108 of the 1976 United States Copyright Ac, without permission of the author.

All inquiries should be addressed to the Math Notion.

ISBN: 978-1-63620-043-9

The Math Notion

Michael Smith has been a math instructor for over a decade now. He launched the Math Notion. Since 2006, we have devoted our time to both teaching and developing exceptional math learning materials. As a test prep company, we have worked with thousands of students. We have used the feedback of our students to develop a unique study program that can be used by students to drastically improve their math scores fast and effectively. We have more than a thousand Math learning books including:

- **SAT Math Prep**
- **ACT Math Prep**
- **SSAT/ISEE Math Prep**
- **Accuplacer Math Prep**
- **Common Core Math Prep**
- **many Math Education Workbooks, Study Guides, Practice and Exercise Books**

As an experienced Math test preparation company, we have helped many students raise their standardized test scores—and attend the colleges of their dreams: We tutor online and in person, we teach students in large groups, and we provide training materials and textbooks through our website and through Amazon.

You can contact us via email at:

info@Mathnotion.com

Common Core Subject Test Mathematics Grade 8

Get the Targeted Practice You Need to Ace the Common Core Math Test!

Common Core Subject Test Mathematics Grade 8 includes easy-to-follow instructions, helpful examples, and plenty of math practice problems to assist students to master each concept, brush up their problem-solving skills, and create confidence.

The Common Core math practice book provides numerous opportunities to evaluate basic skills along with abundant remediation and intervention activities. It is a skill that permits you to quickly master intricate information and produce better leads in less time.

Students can boost their test-taking skills by taking the book's two practice Common Core Math exams. All test questions answered and explained in detail.

Important Features of the 8th grade Common Core Math Book:

- A **complete review** of Common Core math test topics,
- Over 2,500 practice problems covering all topics tested,
- The most important concepts you need to know,
- Clear and concise, easy-to-follow sections,
- Well designed for enhanced learning and interest,
- Hands-on experience with all question types,
- **2 full-length practice tests** with detailed answer explanations,
- Cost-Effective Pricing,

Powerful math exercises to help you avoid traps and pacing yourself to beat the Common Core test. Students will gain valuable experience and raise their confidence by taking 8th grade math practice tests, learning about test structure, and gaining a deeper understanding of what is tested on the Common Core math grade 8. If ever there was a book to respond to the pressure to increase students' test scores, this is it.

Common Core Subject Test Mathematics Grade 8

WWW.MathNotion.COM

… So Much More Online!

- ✓ FREE Math Lessons
- ✓ More Math Learning Books!
- ✓ Mathematics Worksheets
- ✓ Online Math Tutors

For a PDF Version of This Book

Please Visit WWW.MathNotion.com

Common Core Subject Test Mathematics Grade 8

Contents

Chapter 1 : Integers and Number Theory 11
   Rounding 12
   Rounding and Estimates 13
   Adding and Subtracting Integers 14
   Multiplying and Dividing Integers 15
   Order of Operations 16
   Ordering Integers and Numbers 17
   Integers and Absolute Value 18
   Factoring Numbers 19
   Greatest Common Factor 20
   Least Common Multiple 21
   Answers of Worksheets 22

Chapter 2 : Fractions and Decimals 25
   Simplifying Fractions 26
   Adding and Subtracting Fractions 27
   Multiplying and Dividing Fractions 28
   Adding and Subtracting Mixed Numbers 29
   Multiplying and Dividing Mixed Numbers 30
   Adding and Subtracting Decimals 31
   Multiplying and Dividing Decimals 32
   Comparing Decimals 33
   Rounding Decimals 34
   Answers of Worksheets 35

Chapter 3 : Proportions, Ratios, and Percent 38
   Simplifying Ratios 39
   Proportional Ratios 40
   Similarity and Ratios 41
   Ratio and Rates Word Problems 42
   Percentage Calculations 43
   Percent Problems 44
   Discount, Tax and Tip 45
   Percent of Change 46
   Simple Interest 47

Common Core Subject Test Mathematics Grade 8

Answers of Worksheets 48

Chapter 4 : Exponents and Radicals Expressions 51
Multiplication Property of Exponents 52
Zero and Negative Exponents 53
Division Property of Exponents 54
Powers of Products and Quotients 55
Negative Exponents and Negative Bases 56
Scientific Notation 57
Square Roots 58
Simplifying Radical Expressions 59
Multiplying Radical Expressions 60
Simplifying Radical Expressions Involving Fractions 61
Adding and Subtracting Radical Expressions 62
Answers of Worksheets 63

Chapter 5 : Algebraic Expressions 68
Simplifying Variable Expressions 69
Simplifying Polynomial Expressions 70
Translate Phrases into an Algebraic Statement 71
The Distributive Property 72
Evaluating One Variable Expressions 73
Evaluating Two Variables Expressions 74
Combining like Terms 75
Answers of Worksheets 76

Chapter 6 Equations and Inequalities 78
One–Step Equations 79
Multi–Step Equations 80
Graphing Single-Variable Inequalities 81
One–Step Inequalities 82
Multi-Step Inequalities 83
Systems of Equations 84
Systems of Equations Word Problems 85
Answers of Worksheets 86

Chapter 7 : Linear Functions 89
Finding Slope 90
Graphing Lines Using Line Equation 91
Writing Linear Equations 92
Graphing Linear Inequalities 93

Common Core Subject Test Mathematics Grade 8

Finding Midpoint 94
Finding Distance of Two Points 95
Answers of Worksheets 96

Chapter 8 : Polynomials 99
Writing Polynomials in Standard Form 100
Simplifying Polynomials 101
Adding and Subtracting Polynomials 102
Multiplying Monomials 103
Multiplying and Dividing Monomials 104
Multiplying a Polynomial and a Monomial 105
Multiplying Binomials 106
Factoring Trinomials 107
Operations with Polynomials 108
Answers of Worksheets 109

Chapter 9 : Functions Operations and Quadratic 113
Evaluating Function 114
Adding and Subtracting Functions 115
Multiplying and Dividing Functions 116
Composition of Functions 117
Quadratic Equation 118
Solving Quadratic Equations 119
Quadratic Formula and the Discriminant 120
Graphing Quadratic Functions 121
Quadratic Inequalities 122
Domain and Range of Radical Functions 123
Solving Radical Equations 124
Answers of Worksheets 125

Chapter 10 : Geometry and Solid Figures 130
Angles 131
Pythagorean Relationship 132
Triangles 133
Polygons 134
Trapezoids 135
Circles 136
Cubes 137
Rectangular Prism 137
Cylinder 139

Common Core Subject Test Mathematics Grade 8

    Pyramids and Cone..140

    Answers of Worksheets..141

Chapter 11 : Statistics and Probability 143

    Mean and Median 144

    Mode and Range 145

    Times Series 146

    Stem–and–Leaf Plot...147

    Pie Graph 148

    Probability Problems...149

    Factorials 150

    Combinations and Permutations 151

    Answers of Worksheets..152

Chapter 12 : Common Core Math Practice Tests 155

    Common Core Practice Test 1 159

    Common Core Practice Test 2 171

Chapter 13 : Answers and Explanations...185

    Answer Key...185

    Practice Test 1 187

    Practice Test 2 193

Common Core Subject Test Mathematics Grade 8

Chapter 1:
Integers and Number Theory

Topics that you will practice in this chapter:

- ✓ Rounding
- ✓ Whole Number Addition and Subtraction
- ✓ Whole Number Multiplication and Division
- ✓ Rounding and Estimates
- ✓ Adding and Subtracting Integers
- ✓ Multiplying and Dividing Integers
- ✓ Order of Operations
- ✓ Ordering Integers and Numbers
- ✓ Integers and Absolute Value
- ✓ Factoring Numbers
- ✓ Greatest Common Factor (GCF)
- ✓ Least Common Multiple (LCM)

"Wherever there is number, there is beauty." −Proclus

Common Core Subject Test Mathematics Grade 8

Rounding

✎ Round each number to the nearest ten.

1) 42 = ___   5) 19 = ___   9) 48 = ___

2) 88 = ___   6) 25 = ___   10) 81 = ___

3) 24 = ___   7) 93 = ___   11) 58 = ___

4) 57 = ___   8) 71 = ___   12) 87 = ___

✎ Round each number to the nearest hundred.

13) 198 = ___   17) 321 = ___   21) 580 = ___

14) 387 = ___   18) 433 = ___   22) 868 = ___

15) 816 = ___   19) 579 = ___   23) 480 = ___

16) 101 = ___   20) 825 = ___   24) 287 = ___

✎ Round each number to the nearest thousand.

25) 1,382 = ___   29) 9,099 = ___   33) 52,866 = ___

26) 3,420 = ___   30) 22,980 = ___   34) 85,190 = ___

27) 4,254 = ___   31) 45,188 = ___   35) 70,990 = ___

28) 6,861 = ___   32) 16,808 = ___   36) 26,869 = ___

WWW.MathNotion.Com

Common Core Subject Test Mathematics Grade 8

Rounding and Estimates

🔖 **Estimate the sum by rounding each number to the nearest ten.**

1) $13 + 22 = $ _____

2) $71 + 23 = $ _____

3) $61 + 58 = $ _____

4) $56 + 85 = $ _____

5) $368 + 249 = $ _____

6) $330 + 903 = $ _____

7) $471 + 293 = $ _____

8) $1{,}950 + 2{,}655 = $ _____

🔖 **Estimate the product by rounding each number to the nearest ten.**

9) $32 \times 71 = $ _____

10) $12 \times 33 = $ _____

11) $31 \times 83 = $ _____

12) $19 \times 11 = $ _____

13) $42 \times 76 = $ _____

14) $63 \times 34 = $ _____

15) $19 \times 31 = $ _____

16) $59 \times 71 = $ _____

🔖 **Estimate the sum or product by rounding each number to the nearest ten.**

17) $\begin{array}{r} 29 \\ \times\ 12 \\ \hline \end{array}$

18) $\begin{array}{r} 37 \\ \times\ 26 \\ \hline \end{array}$

19) $\begin{array}{r} 48 \\ +\ 82 \\ \hline \end{array}$

20) $\begin{array}{r} 65 \\ +44 \\ \hline \end{array}$

21) $\begin{array}{r} 37 \\ \times\ 14 \\ \hline \end{array}$

22) $\begin{array}{r} 71 \\ +\ 32 \\ \hline \end{array}$

WWW.MathNotion.Com

Adding and Subtracting Integers

✏️ **Find each sum.**

1) $14 + (-6) =$

2) $(-13) + (-20) =$

3) $5 + (-28) =$

4) $50 + (-12) =$

5) $(-7) + (-15) + 3 =$

6) $30 + (-14) + 8 =$

7) $40 + (-10) + (-14) + 17 =$

8) $(-15) + (-20) + 13 + 35 =$

9) $40 + (-20) + (38 - 29) =$

10) $28 + (-12) + (30 - 12) =$

✏️ **Find each difference.**

11) $(-18) - (-7) =$

12) $25 - (-14) =$

13) $(-20) - 36 =$

14) $34 - (-19) =$

15) $51 - (30 - 21) =$

16) $17 - (5) - (-24) =$

17) $(35 + 20) - (-46) =$

18) $48 - 16 - (-8) =$

19) $62 - (28 + 17) - (-15) =$

20) $58 - (-23) - (-31) =$

21) $19 - (-8) - (-13) =$

22) $(19 - 24) - (-14) =$

23) $27 - 33 - (-21) =$

24) $58 - (32 + 24) - (-9) =$

25) $36 - (-30) + (-17) =$

26) $27 - (-42) + (-31) =$

Common Core Subject Test Mathematics Grade 8

Multiplying and Dividing Integers

✎ **Find each product.**

1) $(-9) \times (-5) =$

2) $(-3) \times 9 =$

3) $8 \times (-12) =$

4) $(-7) \times (-20) =$

5) $(-3) \times (-5) \times 6 =$

6) $(14 - 3) \times (-8) =$

7) $12 \times (-9) \times (-3) =$

8) $(140 + 10) \times (-2) =$

9) $10 \times (-12 + 8) \times 3 =$

10) $(-8) \times (-5) \times (-10) =$

✎ **Find each quotient.**

11) $42 \div (-7) =$

12) $(-48) \div (-6) =$

13) $(-40) \div (-8) =$

14) $54 \div (-2) =$

15) $152 \div 19 =$

16) $(-144) \div (-12) =$

17) $180 \div (-10) =$

18) $(-312) \div (-12) =$

19) $221 \div (-13) =$

20) $(-126) \div (6) =$

21) $(-161) \div (-7) =$

22) $-266 \div (-14) =$

23) $(-120) \div (-4) =$

24) $270 \div (-18) =$

25) $(-208) \div (-8) =$

26) $(135) \div (-15) =$

Order of Operations

✏️ **Evaluate each expression.**

1) $7 + (5 \times 4) =$

2) $14 - (3 \times 6) =$

3) $(19 \times 4) + 16 =$

4) $(16 - 7) - (8 \times 2) =$

5) $27 + (18 \div 3) =$

6) $(18 \times 8) \div 6 =$

7) $(32 \div 4) \times (-2) =$

8) $(9 \times 4) + (32 - 18) =$

9) $24 + (4 \times 3) + 7 =$

10) $(36 \times 3) \div (2 + 2) =$

11) $(-7) + (12 \times 3) + 11 =$

12) $(8 \times 5) - (24 \div 6) =$

13) $(7 \times 6 \div 3) - (12 + 9) =$

14) $(13 + 5 - 14) \times 3 - 2 =$

15) $(20 - 14 + 30) \times (64 \div 4) =$

16) $32 + (28 - (36 \div 9)) =$

17) $(7 + 6 - 4 - 7) + (15 \div 5) =$

18) $(85 - 20) + (20 - 18 + 7) =$

19) $(20 \times 2) + (14 \times 3) - 22 =$

20) $18 + 5 - (30 \times 3) + 20 =$

Common Core Subject Test Mathematics Grade 8

Ordering Integers and Numbers

✎ **Order each set of integers from least to greatest.**

1) $8, -10, -5, -3, 4$ ___, ___, ___, ___, ___, ___

2) $-10, -18, 6, 14, 27$ ___, ___, ___, ___, ___, ___

3) $15, -8, -21, 21, -23$ ___, ___, ___, ___, ___, ___

4) $-14, -40, 23, -12, 47$ ___, ___, ___, ___, ___, ___

5) $59, -54, 32, -57, 36$ ___, ___, ___, ___, ___, ___

6) $68, 26, -19, 47, -34$ ___, ___, ___, ___, ___, ___

✎ **Order each set of integers from greatest to least.**

7) $18, 36, -16, -18, -10$ ___, ___, ___, ___, ___, ___

8) $27, 34, -12, -24, 94$ ___, ___, ___, ___, ___, ___

9) $50, -21, -13, 42, -2$ ___, ___, ___, ___, ___, ___

10) $37, 46, -20, -16, 86$ ___, ___, ___, ___, ___, ___

11) $-18, 88, -26, -59, 75$ ___, ___, ___, ___, ___, ___

12) $-65, -30, -25, 3, 14$ ___, ___, ___, ___, ___, ___

Common Core Subject Test Mathematics Grade 8

Integers and Absolute Value

✎ **Write absolute value of each number.**

1) $|-2| =$

2) $|-27| =$

3) $|-20| =$

4) $|14| =$

5) $|6| =$

6) $|-55| =$

7) $|16| =$

8) $|2| =$

9) $|54| =$

10) $|-4| =$

11) $|-11|$

12) $|88| =$

13) $|0| =$

14) $|79| =$

15) $|-32| =$

16) $|-17| =$

17) $|42| =$

18) $|-46| =$

19) $|1| =$

20) $|-40| =$

✎ **Evaluate the value.**

21) $|-5| - \frac{|-21|}{7} =$

22) $14 - |3 - 15| - |-4| =$

23) $\frac{|-32|}{4} \times |-4| =$

24) $\frac{|7 \times (-3)|}{7} \times \frac{|-19|}{3} =$

25) $|4 \times (-5)| + \frac{|-40|}{5} =$

26) $\frac{|-45|}{9} \times \frac{|-24|}{12} =$

27) $|-12 + 8| \times \frac{|-7 \times 7|}{7}$

28) $\frac{|-11 \times 2|}{4} \times |-16| =$

WWW.MathNotion.Com

Factoring Numbers

✏️ **List all positive factors of each number.**

1) 9

2) 16

3) 24

4) 30

5) 26

6) 46

7) 20

8) 68

9) 28

10) 98

11) 14

12) 54

13) 55

14) 18

15) 63

16) 34

17) 50

18) 62

19) 95

20) 64

21) 70

22) 45

23) 22

24) 65

Common Core Subject Test Mathematics Grade 8

Greatest Common Factor

✏️ **Find the GCF for each number pair.**

1) 6, 2

2) 4, 5

3) 3, 12

4) 7, 3

5) 5, 10

6) 8, 48

7) 6, 18

8) 9, 15

9) 12, 18

10) 4, 36

11) 6, 10

12) 28, 52

13) 25, 10

14) 22, 24

15) 9, 54

16) 8, 54

17) 42, 14

18) 16, 40

19) 9, 2, 3

20) 5, 15, 10

21) 7, 9, 2

22) 16, 64

23) 30, 48

24) 36, 63

Common Core Subject Test Mathematics Grade 8

Least Common Multiple

✎ **Find the LCM for each number pair.**

1) 6, 9

2) 15, 45

3) 16, 40

4) 12, 36

5) 18, 27

6) 14, 42

7) 6, 30

8) 8, 56

9) 7, 21

10) 8, 20

11) 15, 25

12) 7, 9

13) 4, 11

14) 8, 28

15) 28, 56

16) 40, 50

17) 12, 13

18) 22, 11

19) 36, 20

20) 15, 35

21) 18, 81

22) 30, 54

23) 18, 45

24) 75, 25

Common Core Subject Test Mathematics Grade 8

Answers of Worksheets

Rounding

1) 40
2) 90
3) 20
4) 60
5) 20
6) 30
7) 90
8) 70
9) 50
10) 80
11) 60
12) 90
13) 200
14) 400
15) 800
16) 100
17) 300
18) 400
19) 600
20) 800
21) 600
22) 900
23) 500
24) 300
25) 1,000
26) 3,000
27) 4,000
28) 7,000
29) 9,000
30) 23,000
31) 45,000
32) 17,000
33) 53,000
34) 85,000
35) 71,000
36) 27,000

Rounding and Estimates

1) 30
2) 90
3) 120
4) 150
5) 620
6) 1,230
7) 760
8) 4,610
9) 2,100
10) 300
11) 2,400
12) 200
13) 3,200
14) 1,800
15) 600
16) 4,200
17) 300
18) 1,200
19) 130
20) 110
21) 400
22) 100

Adding and Subtracting Integers

1) 8
2) −33
3) −23
4) 38
5) −19
6) 24
7) 33
8) 13
9) 29
10) 34
11) −11
12) 39
13) −56
14) 53
15) 42
16) 36
17) 101
18) 40
19) 32
20) 112
21) 40
22) 9
23) 15
24) 11
25) 49
26) 38

Multiplying and Dividing Integers

1) 45
2) −27
3) −96
4) 140
5) 90
6) −88
7) 324
8) −300
9) −120
10) −400
11) −6
12) 8
13) 5
14) −27
15) 8
16) 12
17) −18
18) 26
19) −17
20) −21

Common Core Subject Test Mathematics Grade 8

21) 23   23) 30   25) 26
22) 19   24) −15   26) −9

Order of Operations

1) 27   6) 24   11) 40   16) 56
2) −4   7) −16   12) 36   17) 5
3) 92   8) 50   13) −7   18) 74
4) −7   9) 43   14) 10   19) 60
5) 33   10) 27   15) 576   20) −47

Ordering Integers and Numbers

1) −10, −5, −3, 4, 8
2) −18, −10, 6, 14, 27
3) −23, −21, −8, 15, 21
4) −40, −14, −12, 23, 47
5) −57, −54, 32, 36, 59
6) −34, −19, 26, 47, 68
7) 36, 18, −10, −16, −18
8) 94, 34, 27, −12, −24
9) 50, 42, −2, −13, −21
10) 86, 46, 37, −16, −20
11) 88, 75, −18, −26, −59
12) 14, 3, −25, −30, −65

Integers and Absolute Value

1) 2   8) 2   15) 32   22) −2
2) 27   9) 54   16) 17   23) 32
3) 20   10) 4   17) 42   24) 19
4) 14   11) 11   18) 46   25) 28
5) 6   12) 88   19) 1   26) 10
6) 55   13) 0   20) 40   27) 28
7) 16   14) 79   21) 2   28) 88

Factoring Numbers

1) 1, 3, 9
2) 1, 2, 4, 8, 16
3) 1, 2, 3, 4, 6, 8, 12, 24
4) 1, 2, 3, 5, 6, 10, 15, 30
5) 1, 2, 13, 26
6) 1, 2, 23, 46
7) 1, 2, 4, 5, 10, 20
8) 1, 2, 4, 17, 34, 68
9) 1, 2, 4, 7, 14, 28
10) 1, 2, 7, 14, 49, 98
11) 1, 2, 7, 14
12) 1, 2, 3, 6, 9, 18, 27, 54
13) 1, 5, 11, 55
14) 1, 2, 3, 6, 9, 18
15) 1, 3, 7, 9, 21, 63
16) 1, 2, 17, 34
17) 1, 2, 5, 10, 25, 50
18) 1, 2, 31, 62
19) 1, 5, 19, 95
20) 1, 2, 4, 8, 16, 32, 64
21) 1, 2, 5, 7, 10, 14, 35, 70
22) 1, 3, 5, 9, 15, 45
23) 1, 2, 11, 22
24) 1, 5, 13, 65

Common Core Subject Test Mathematics Grade 8

Greatest Common Factor

1) 2	7) 6	13) 5	19) 1
2) 1	8) 3	14) 2	20) 5
3) 3	9) 6	15) 9	21) 1
4) 1	10) 4	16) 2	22) 16
5) 5	11) 2	17) 14	23) 6
6) 8	12) 4	18) 8	24) 9

Least Common Multiple

1) 18	7) 30	13) 44	19) 180
2) 45	8) 56	14) 56	20) 105
3) 80	9) 21	15) 56	21) 162
4) 36	10) 40	16) 200	22) 270
5) 54	11) 75	17) 156	23) 90
6) 42	12) 63	18) 22	24) 75

Common Core Subject Test Mathematics Grade 8

Chapter 2:
Fractions and Decimals

Topics that you will practice in this chapter:

- ✓ Simplifying Fractions
- ✓ Adding and Subtracting Fractions
- ✓ Multiplying and Dividing Fractions
- ✓ Adding and Subtract Mixed Numbers
- ✓ Multiplying and Dividing Mixed Numbers
- ✓ Adding and Subtracting Decimals
- ✓ Multiplying and Dividing Decimals
- ✓ Comparing Decimals
- ✓ Rounding Decimals

"A Man is like a fraction whose numerator is what he is and whose denominator is what he thinks of himself. The larger the denominator, the smaller the fraction." –Tolstoy

Common Core Subject Test Mathematics Grade 8

Simplifying Fractions

✎ Simplify each fraction to its lowest terms.

1) $\dfrac{5}{10} =$

2) $\dfrac{28}{35} =$

3) $\dfrac{27}{36} =$

4) $\dfrac{40}{80} =$

5) $\dfrac{14}{56} =$

6) $\dfrac{32}{48} =$

7) $\dfrac{52}{65} =$

8) $\dfrac{15}{60} =$

9) $\dfrac{80}{160} =$

10) $\dfrac{55}{77} =$

11) $\dfrac{28}{112} =$

12) $\dfrac{32}{64} =$

13) $\dfrac{63}{72} =$

14) $\dfrac{81}{90} =$

15) $\dfrac{35}{105} =$

16) $\dfrac{25}{70} =$

17) $\dfrac{80}{280} =$

18) $\dfrac{12}{81} =$

19) $\dfrac{36}{186} =$

20) $\dfrac{240}{540} =$

21) $\dfrac{70}{560} =$

✎ Find the answer for each problem.

22) Which of the following fractions equal to $\dfrac{3}{4}$? ____

　A. $\dfrac{60}{90}$ B. $\dfrac{43}{104}$ C. $\dfrac{48}{64}$ D. $\dfrac{150}{300}$

23) Which of the following fractions equal to $\dfrac{5}{8}$? ____

　A. $\dfrac{125}{200}$ B. $\dfrac{115}{200}$ C. $\dfrac{50}{100}$ D. $\dfrac{30}{90}$

24) Which of the following fractions equal to $\dfrac{3}{7}$? ____

　A. $\dfrac{58}{116}$ B. $\dfrac{54}{126}$ C. $\dfrac{270}{167}$ D. $\dfrac{42}{63}$

WWW.MathNotion.Com

Adding and Subtracting Fractions

✏️ **Find the sum.**

1) $\dfrac{5}{9} + \dfrac{4}{9} =$

2) $\dfrac{1}{2} + \dfrac{1}{7} =$

3) $\dfrac{3}{8} + \dfrac{1}{4} =$

4) $\dfrac{3}{5} + \dfrac{1}{2} =$

5) $\dfrac{1}{4} + \dfrac{3}{5} =$

6) $\dfrac{7}{8} + \dfrac{3}{8} =$

7) $\dfrac{1}{2} + \dfrac{7}{10} =$

8) $\dfrac{2}{5} + \dfrac{2}{3} =$

9) $\dfrac{5}{7} + \dfrac{2}{3} =$

10) $\dfrac{7}{12} + \dfrac{3}{4} =$

11) $\dfrac{5}{6} + \dfrac{2}{5} =$

12) $\dfrac{1}{12} + \dfrac{2}{3} =$

✏️ **Find the difference.**

13) $\dfrac{1}{3} - \dfrac{1}{6} =$

14) $\dfrac{3}{4} - \dfrac{1}{8} =$

15) $\dfrac{1}{2} - \dfrac{1}{3} =$

16) $\dfrac{1}{4} - \dfrac{1}{5} =$

17) $\dfrac{5}{8} - \dfrac{2}{3} =$

18) $\dfrac{1}{4} - \dfrac{1}{7} =$

19) $\dfrac{5}{6} - \dfrac{1}{9} =$

20) $\dfrac{3}{4} - \dfrac{1}{6} =$

21) $\dfrac{7}{8} - \dfrac{1}{12} =$

22) $\dfrac{8}{15} - \dfrac{3}{5} =$

23) $\dfrac{3}{12} - \dfrac{1}{14} =$

24) $\dfrac{10}{13} - \dfrac{7}{26} =$

25) $\dfrac{6}{7} - \dfrac{3}{4} =$

26) $\dfrac{4}{5} - \dfrac{1}{8} =$

27) $\dfrac{4}{7} - \dfrac{2}{35} =$

28) $\dfrac{9}{16} - \dfrac{2}{8} =$

29) $\dfrac{8}{9} - \dfrac{7}{18} =$

30) $\dfrac{1}{2} - \dfrac{4}{9} =$

Common Core Subject Test Mathematics Grade 8

Multiplying and Dividing Fractions

✎ Find the value of each expression in lowest terms.

1) $\frac{1}{5} \times \frac{15}{5} =$

2) $\frac{9}{12} \times \frac{4}{9} =$

3) $\frac{1}{16} \times \frac{8}{10} =$

4) $\frac{1}{24} \times \frac{8}{10} =$

5) $\frac{1}{5} \times \frac{1}{4} =$

6) $\frac{7}{9} \times \frac{1}{7} =$

7) $\frac{6}{7} \times \frac{1}{3} =$

8) $\frac{2}{8} \times \frac{2}{8} =$

9) $\frac{5}{8} \times \frac{3}{5} =$

10) $\frac{4}{7} \times \frac{1}{8} =$

11) $\frac{7}{15} \times \frac{5}{7} =$

12) $\frac{3}{10} \times \frac{5}{9} =$

✎ Find the value of each expression in lowest terms.

13) $\frac{1}{4} \div \frac{1}{8} =$

14) $\frac{1}{10} \div \frac{1}{5} =$

15) $\frac{3}{4} \div \frac{1}{5} =$

16) $\frac{1}{3} \div \frac{5}{6} =$

17) $\frac{1}{7} \div \frac{8}{42} =$

18) $\frac{3}{4} \div \frac{1}{6} =$

19) $\frac{2}{7} \div \frac{7}{13} =$

20) $\frac{1}{24} \div \frac{3}{16} =$

21) $\frac{7}{12} \div \frac{5}{6} =$

22) $\frac{22}{18} \div \frac{11}{9} =$

23) $\frac{9}{35} \div \frac{3}{7} =$

24) $\frac{2}{7} \div \frac{8}{21} =$

25) $\frac{1}{9} \div \frac{2}{5} =$

26) $\frac{5}{12} \div \frac{3}{5} =$

27) $\frac{3}{20} \div \frac{1}{6} =$

28) $\frac{8}{20} \div \frac{3}{4} =$

29) $\frac{5}{6} \div \frac{2}{9} =$

30) $\frac{5}{11} \div \frac{3}{4} =$

WWW.MathNotion.Com

Common Core Subject Test Mathematics Grade 8

Adding and Subtracting Mixed Numbers

✎ Find the sum.

1) $3\frac{1}{3} + 2\frac{1}{6} =$

2) $4\frac{1}{2} + 3\frac{1}{2} =$

3) $3\frac{3}{8} + 1\frac{1}{8} =$

4) $2\frac{1}{4} + 2\frac{1}{3} =$

5) $3\frac{5}{6} + 2\frac{7}{12} =$

6) $5\frac{4}{15} + 3\frac{3}{5} =$

7) $2\frac{1}{3} + 4\frac{3}{7} =$

8) $3\frac{1}{2} + 4\frac{2}{5} =$

9) $5\frac{2}{5} + 6\frac{3}{7} =$

10) $8\frac{5}{16} + 6\frac{1}{12} =$

✎ Find the difference.

11) $3\frac{1}{4} - 1\frac{3}{4} =$

12) $6\frac{3}{5} - 4\frac{2}{5} =$

13) $4\frac{1}{3} - 3\frac{1}{9} =$

14) $7\frac{1}{7} - 5\frac{1}{2} =$

15) $5\frac{1}{3} - 2\frac{1}{12} =$

16) $8\frac{1}{5} - 4\frac{1}{3} =$

17) $9\frac{1}{4} - 6\frac{1}{8} =$

18) $11\frac{7}{15} - 8\frac{3}{5} =$

19) $14\frac{5}{6} - 11\frac{3}{5} =$

20) $18\frac{2}{7} - 14\frac{1}{5} =$

21) $9\frac{1}{3} - 4\frac{1}{4} =$

22) $6\frac{1}{8} - 4\frac{1}{16} =$

23) $19\frac{3}{8} - 15\frac{1}{3} =$

24) $11\frac{1}{9} - 8\frac{1}{8} =$

25) $17\frac{1}{7} - 11\frac{1}{5} =$

26) $16\frac{2}{9} - 9\frac{5}{7} =$

Common Core Subject Test Mathematics Grade 8

Multiplying and Dividing Mixed Numbers

✏ **Find the product.**

1) $5\frac{1}{2} \times 2\frac{1}{4} =$

2) $5\frac{1}{3} \times 4\frac{1}{3} =$

3) $5\frac{3}{4} \times 6\frac{1}{4} =$

4) $3\frac{1}{3} \times 2\frac{3}{5} =$

5) $4\frac{8}{10} \times 1\frac{1}{24} =$

6) $6\frac{2}{7} \times 1\frac{1}{11} =$

7) $8\frac{2}{3} \times 3\frac{1}{2} =$

8) $3\frac{4}{7} \times 2\frac{1}{5} =$

9) $5\frac{2}{8} \times 4\frac{1}{6} =$

10) $7\frac{3}{3} \times 1\frac{3}{8} =$

✏ **Find the quotient.**

11) $2\frac{2}{5} \div 4\frac{1}{5} =$

12) $4\frac{1}{6} \div 3\frac{1}{3} =$

13) $6\frac{1}{3} \div 1\frac{1}{2} =$

14) $7\frac{1}{10} \div 2\frac{2}{5} =$

15) $3\frac{1}{3} \div 1\frac{1}{9} =$

16) $1\frac{1}{10} \div 4\frac{1}{2} =$

17) $1\frac{3}{16} \div 5\frac{1}{4} =$

18) $4\frac{1}{3} \div 4\frac{3}{4} =$

19) $9\frac{1}{3} \div 2\frac{1}{4} =$

20) $15\frac{1}{3} \div 5\frac{1}{2} =$

21) $4\frac{1}{6} \div 1\frac{1}{5} =$

22) $1\frac{1}{18} \div 1\frac{2}{9} =$

23) $4\frac{2}{7} \div 1\frac{3}{10} =$

24) $7\frac{1}{3} \div 2\frac{2}{11} =$

25) $8\frac{2}{5} \div 1\frac{1}{6} =$

26) $9\frac{1}{3} \div 2\frac{1}{7} =$

WWW.MathNotion.Com

Common Core Subject Test Mathematics Grade 8

Adding and Subtracting Decimals

✎ **Add and subtract decimals.**

1) 35.19 − 24.28 = _____

4) 38.72 − 21.68 = _____

7) 86.09 − 35.14 = _____

2) 34.29 + 42.58 = _____

5) 57.39 + 26.54 = _____

8) 54.51 + 32.66 = _____

3) 61.20 + 33.75 = _____

6) 70.24 − 42.35 = _____

9) 114.21 − 88.69 = _____

✎ **Find the missing number.**

10) ___ + 2.8 = 5.4

11) 4.1 + ___ = 5.88

12) 6.45 + ___ = 8

13) 7.25 − ___ = 3.40

14) ___ − 2.35 = 4.25

15) ___ − 19.85 = 6.54

16) 22.15 + ___ = 28.95

17) ___ − 37.16 = 9.42

18) ___ + 24.50 = 34.19

19) 72.40 + ___ = 125.20

Multiplying and Dividing Decimals

✏ **Find the product.**

1) 0.5 × 0.6 =

2) 3.3 × 0.4 =

3) 1.28 × 0.5 =

4) 0.35 × 0.6 =

5) 1.85 × 0.6 =

6) 0.24 × 0.5 =

7) 5.25 × 1.4 =

8) 18.5 × 4.6 =

9) 15.4 × 6.8 =

10) 19.5 × 2.6 =

11) 32.2 × 1.5 =

12) 78.4 × 4.5 =

✏ **Find the quotient.**

13) 1.85 ÷ 10 =

14) 74.6 ÷ 100 =

15) 3.6 ÷ 3 =

16) 9.6 ÷ 0.4 =

17) 15.5 ÷ 0.5 =

18) 32.8 ÷ 0.2 =

19) 22.15 ÷ 1,000 =

20) 53.55 ÷ 0.7 =

21) 322.2 ÷ 0.2 =

22) 50.67 ÷ 0.18 =

23) 77.4 ÷ 0.8 =

24) 27.93 ÷ 0.03 =

Common Core Subject Test Mathematics Grade 8

Comparing Decimals

✎ **Write the correct comparison symbol (>, < or =).**

1) 0.70 ☐ 0.070

2) 0.049 ☐ 0.49

3) 5.090 ☐ 5.09

4) 2.57 ☐ 2.05

5) 9.03 ☐ 0.930

6) 6.06 ☐ 6.6

7) 7.02 ☐ 7.020

8) 3.04 ☐ 3.2

9) 3.61 ☐ 3.245

10) 0.986 ☐ 0.0986

11) 17.24 ☐ 17.240

12) 0.759 ☐ 0.81

13) 9.040 ☐ 9.40

14) 5.73 ☐ 5.213

15) 9.44 ☐ 9.404

16) 7.17 ☐ 7.170

17) 4.85 ☐ 4.085

18) 9.041 ☐ 9.40

19) 3.033 ☐ 3.030

20) 4.97 ☐ 4.970

Rounding Decimals

✎ Round each decimal to the nearest whole number.

1) 28.12 3) 16.22 5) 7.95

2) 6.9 4) 8.5 6) 52.7

✎ Round each decimal to the nearest tenth.

7) 31.761 9) 94.729 11) 13.219

8) 14.421 10) 77.89 12) 59.89

✎ Round each decimal to the nearest hundredth.

13) 8.428 15) 55.3786 17) 62.241

14) 23.812 16) 231.912 18) 19.447

✎ Round each decimal to the nearest thousandth.

19) 15.54324 21) 243.8652 23) 67.1983

20) 34.62586 22) 80.4529 24) 72.36788

Common Core Subject Test Mathematics Grade 8

Answers of Worksheets

Simplifying Fractions

1) $\frac{1}{2}$
2) $\frac{4}{5}$
3) $\frac{3}{4}$
4) $\frac{1}{2}$
5) $\frac{1}{4}$
6) $\frac{2}{3}$
7) $\frac{4}{5}$
8) $\frac{1}{4}$
9) $\frac{1}{2}$
10) $\frac{5}{7}$
11) $\frac{1}{4}$
12) $\frac{1}{2}$
13) $\frac{7}{8}$
14) $\frac{9}{10}$
15) $\frac{1}{3}$
16) $\frac{5}{14}$
17) $\frac{2}{7}$
18) $\frac{4}{27}$
19) $\frac{6}{31}$
20) $\frac{4}{9}$
21) $\frac{1}{8}$
22) C
23) A
24) B

Adding and Subtracting Fractions

1) $\frac{9}{9} = 1$
2) $\frac{9}{14}$
3) $\frac{5}{8}$
4) $1\frac{1}{10}$
5) $\frac{17}{20}$
6) $1\frac{1}{4}$
7) $1\frac{1}{5}$
8) $1\frac{1}{15}$
9) $1\frac{8}{21}$
10) $1\frac{1}{3}$
11) $1\frac{7}{30}$
12) $\frac{3}{4}$
13) $\frac{1}{6}$
14) $\frac{5}{8}$
15) $\frac{1}{6}$
16) $\frac{1}{20}$
17) $-\frac{1}{24}$
18) $\frac{3}{28}$
19) $\frac{13}{18}$
20) $\frac{7}{12}$
21) $\frac{19}{24}$
22) $-\frac{1}{15}$
23) $\frac{5}{28}$
24) $\frac{1}{2}$
25) $\frac{3}{28}$
26) $\frac{27}{40}$
27) $\frac{18}{35}$
28) $\frac{5}{16}$
29) $\frac{1}{2}$
30) $\frac{1}{18}$

Multiplying and Dividing Fractions

1) $\frac{3}{5}$
2) $\frac{1}{3}$
3) $\frac{1}{20}$
4) $\frac{1}{30}$
5) $\frac{1}{20}$
6) $\frac{1}{9}$
7) $\frac{2}{7}$
8) $\frac{1}{16}$
9) $\frac{3}{8}$
10) $\frac{1}{14}$
11) $\frac{1}{3}$
12) $\frac{1}{6}$
13) 2
14) $\frac{1}{2}$
15) $3\frac{3}{4}$
16) $\frac{2}{5}$

Common Core Subject Test Mathematics Grade 8

17) $\frac{3}{4}$

18) $4\frac{1}{2}$

19) $\frac{26}{49}$

20) $\frac{2}{9}$

21) $\frac{7}{10}$

22) 1

23) $\frac{3}{5}$

24) $\frac{3}{4}$

25) $\frac{5}{18}$

26) $\frac{25}{36}$

27) $\frac{9}{10}$

28) $\frac{8}{15}$

29) $3\frac{3}{4}$

30) $\frac{20}{33}$

Adding and Subtracting Mixed Numbers

1) $5\frac{1}{2}$

2) 8

3) $4\frac{1}{2}$

4) $4\frac{7}{12}$

5) $6\frac{5}{12}$

6) $8\frac{13}{15}$

7) $6\frac{16}{21}$

8) $7\frac{9}{10}$

9) $11\frac{29}{35}$

10) $14\frac{19}{48}$

11) $1\frac{1}{2}$

12) $2\frac{1}{5}$

13) $1\frac{2}{9}$

14) $1\frac{9}{14}$

15) $3\frac{1}{4}$

16) $3\frac{13}{15}$

17) $3\frac{1}{8}$

18) $2\frac{13}{15}$

19) $3\frac{7}{30}$

20) $4\frac{3}{35}$

21) $5\frac{1}{12}$

22) $2\frac{1}{16}$

23) $4\frac{1}{24}$

24) $2\frac{71}{72}$

25) $5\frac{33}{35}$

26) $6\frac{32}{63}$

Multiplying and Dividing Mixed Numbers

1) $12\frac{3}{8}$

2) $23\frac{1}{9}$

3) $35\frac{15}{16}$

4) $8\frac{2}{3}$

5) 5

6) $6\frac{6}{7}$

7) $30\frac{1}{3}$

8) $7\frac{6}{7}$

9) $21\frac{7}{8}$

10) 11

11) $\frac{4}{7}$

12) $1\frac{1}{4}$

13) $4\frac{2}{9}$

14) $2\frac{23}{24}$

15) 3

16) $\frac{11}{45}$

17) $\frac{19}{84}$

18) $\frac{52}{57}$

19) $4\frac{4}{27}$

20) $2\frac{26}{33}$

21) $3\frac{17}{36}$

22) $\frac{19}{22}$

23) $3\frac{27}{91}$

24) $3\frac{13}{36}$

25) $7\frac{1}{5}$

26) $4\frac{16}{45}$

Adding and Subtracting Decimals

1) 10.91

2) 76.87

3) 94.95

4) 17.04

Common Core Subject Test Mathematics Grade 8

5) 83.93
6) 27.89
7) 50.95
8) 87.17

9) 25.52
10) 2.6
11) 1.78
12) 1.55

13) 3.85
14) 6.6
15) 26.39
16) 6.8

17) 46.58
18) 9.69
19) 52.8

Multiplying and Dividing Decimals

1) 0.3
2) 1.32
3) 0.64
4) 0.21
5) 1.11
6) 0.12

7) 7.35
8) 85.1
9) 104.72
10) 50.7
11) 48.3
12) 352.8

13) 0.185
14) 0.746
15) 1.2
16) 24
17) 31
18) 164

19) 0.02215
20) 76.5
21) 1,611
22) 281.5
23) 96.75
24) 931

Comparing Decimals

1) >
2) <
3) =
4) >
5) >

6) <
7) =
8) <
9) >
10) >

11) =
12) <
13) <
14) >
15) >

16) =
17) >
18) <
19) >
20) =

Rounding Decimals

1) 28
2) 7
3) 16
4) 9
5) 8
6) 53
7) 31.8
8) 14.4

9) 94.7
10) 77.9
11) 13.2
12) 59.9
13) 8.43
14) 23.81
15) 55.38
16) 231.91

17) 62.24
18) 19.45
19) 15.543
20) 34.626
21) 243.865
22) 80.453
23) 67.198
24) 72.368

Common Core Subject Test Mathematics Grade 8

Chapter 3:
Proportions, Ratios, and Percent

Topics that you will practice in this chapter:

- ✓ Simplifying Ratios
- ✓ Proportional Ratios
- ✓ Similarity and Ratios
- ✓ Ratio and Rates Word Problems
- ✓ Percentage Calculations
- ✓ Percent Problems
- ✓ Discount, Tax and Tip
- ✓ Percent of Change
- ✓ Simple Interest

Without mathematics, there's nothing you can do. Everything around you is mathematics. Everything around you is numbers." – Shakuntala Devi

Simplifying Ratios

✍ **Reduce each ratio.**

1) 15: 20 = ___ : ___
2) 7: 70 = ___ : ___
3) 16: 28 = ___ : ___
4) 7: 21 = ___ : ___
5) 4: 40 = ___ : ___
6) 6: 48 = ___ : ___
7) 16: 64 = ___ : ___
8) 10: 25 = ___ : ___

9) 8: 48 = ___ : ___
10) 49: 63 = ___ : ___
11) 18: 27 = ___ : ___
12) 35: 10 = ___ : ___
13) 90: 9 = ___ : ___
14) 24: 32 = ___ : ___
15) 7: 56 = ___ : ___
16) 45: 63 = ___ : ___

17) 56: 72 = ___ : ___
18) 26: 13 = ___ : ___
19) 15: 45 = ___ : ___
20) 28: 4 = ___ : ___
21) 24: 48 = ___ : ___
22) 30: 24 = ___ : ___
23) 70: 140 = ___ : ___
24) 6: 180 = ___ : ___

✍ **Write each ratio as a fraction in simplest form.**

25) 6: 12 =
26) 30: 50 =
27) 15: 35 =
28) 9: 27 =
29) 8: 24 =
30) 18: 84 =
31) 7: 14 =

32) 7: 35 =
33) 40: 96 =
34) 12: 54 =
35) 44: 52 =
36) 12: 27 =
37) 15: 180 =
38) 39: 143 =

39) 20: 300 =
40) 30: 120 =
41) 56: 42 =
42) 26: 130 =
43) 66: 123 =
44) 70: 630 =
45) 75: 125 =

Common Core Subject Test Mathematics Grade 8

Proportional Ratios

✏️ **Fill in the blanks; Calculate each proportion.**

1) $3:8 = \_\_:48$

2) $2:5 = 20:\_\_$

3) $1:9 = \_\_:81$

4) $6:7 = 12:\_\_$

5) $9:2 = 63:\_\_$

6) $8:7 = \_\_:49$

7) $20:3 = \_\_:15$

8) $1:3 = \_\_:75$

9) $7:6 = \_\_:60$

10) $8:5 = \_\_:45$

11) $3:10 = 60:\_\_$

12) $6:11 = 42:\_\_$

✏️ **State if each pair of ratios form a proportion.**

13) $\frac{3}{20}$ and $\frac{9}{60}$

14) $\frac{1}{7}$ and $\frac{6}{42}$

15) $\frac{3}{7}$ and $\frac{24}{56}$

16) $\frac{4}{9}$ and $\frac{12}{18}$

17) $\frac{1}{9}$ and $\frac{12}{81}$

18) $\frac{7}{8}$ and $\frac{21}{28}$

19) $\frac{9}{13}$ and $\frac{27}{39}$

20) $\frac{1}{8}$ and $\frac{8}{64}$

21) $\frac{6}{19}$ and $\frac{30}{85}$

22) $\frac{5}{9}$ and $\frac{40}{81}$

23) $\frac{9}{14}$ and $\frac{108}{168}$

24) $\frac{15}{23}$ and $\frac{360}{552}$

✏️ **Calculate each proportion.**

25) $\frac{20}{25} = \frac{32}{x}, x = \_\_\_\_$

26) $\frac{1}{8} = \frac{32}{x}, x = \_\_\_\_$

27) $\frac{15}{5} = \frac{21}{x}, x = \_\_\_\_$

28) $\frac{1}{7} = \frac{x}{294}, x = \_\_\_\_$

29) $\frac{7}{9} = \frac{x}{81}, x = \_\_\_\_$

30) $\frac{1}{5} = \frac{13}{x}, x = \_\_\_\_$

31) $\frac{9}{5} = \frac{36}{x}, x = \_\_\_\_$

32) $\frac{6}{13} = \frac{48}{x}, x = \_\_\_\_$

33) $\frac{5}{8} = \frac{x}{88}, x = \_\_\_\_$

34) $\frac{4}{15} = \frac{x}{240}, x = \_\_\_\_$

35) $\frac{9}{19} = \frac{x}{266}, x = \_\_\_\_$

36) $\frac{7}{15} = \frac{x}{270}, x = \_\_\_\_$

Common Core Subject Test Mathematics Grade 8

Similarity and Ratios

✎ **Each pair of figures is similar. Find the missing side.**

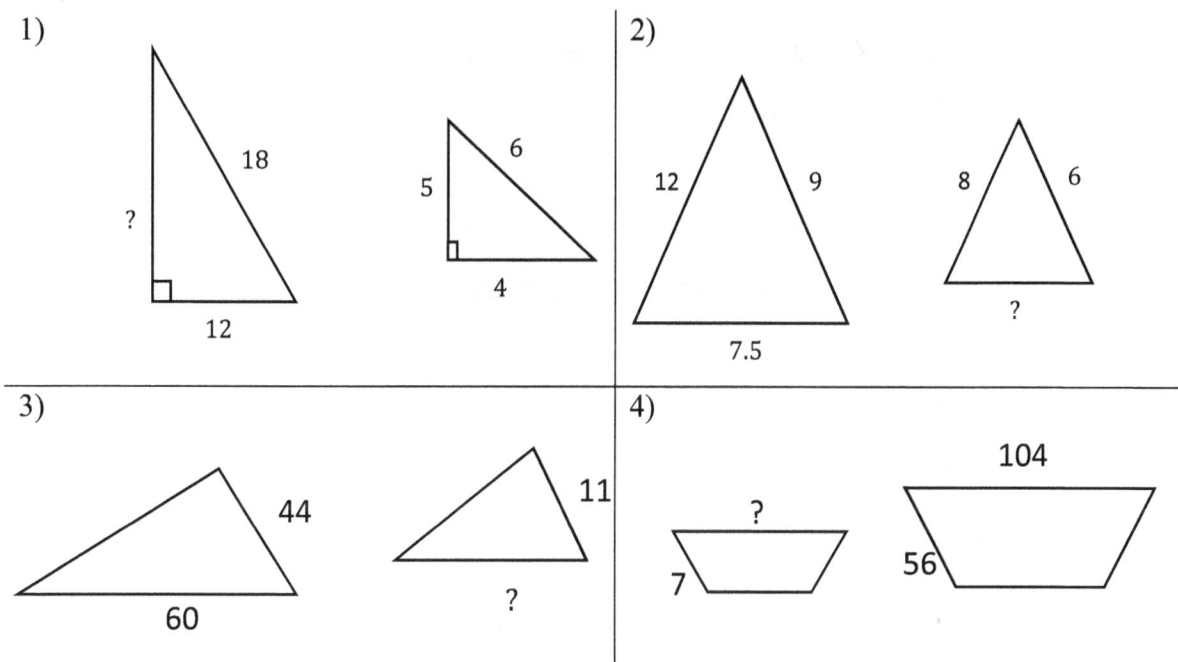

✎ **Calculate.**

5) Two rectangles are similar. The first is 24 feet wide and 120 feet long. The second is 30 feet wide. What is the length of the second rectangle? _____

6) Two rectangles are similar. One is 5 meters by 36 meters. The longer side of the second rectangle is 90 meters. What is the other side of the second rectangle? _____

7) A building casts a shadow 25 ft long. At the same time a girl 10 ft tall casts a shadow 5 ft long. How tall is the building? _____

8) The scale of a map of Texas is 4 inches: 32 miles. If you measure the distance from Dallas to Martin County as 38.4 inches, approximately how far is Martin County from Dallas? _____

WWW.MathNotion.Com

Common Core Subject Test Mathematics Grade 8

Ratio and Rates Word Problems

✍ Find the answer for each word problem.

1) Mason has 24 red cards and 36 green cards. What is the ratio of Mason's red cards to his green cards? _____

2) In a party, 45 soft drinks are required for every 54 guests. If there are 378 guests, how many soft drinks is required? _____

3) In Mason's class, 42 of the students are tall and 24 are short. In Michael's class 84 students are tall and 48 students are short. Which class has a higher ratio of tall to short students? _____

4) The price of 5 apples at the Quick Market is $4.6. The price of 7 of the same apples at Walmart is $5.95. Which place is the better buy? _____

5) The bakers at a Bakery can make 90 bagels in 3 hours. How many bagels can they bake in 24 hours? What is that rate per hour? _____

6) You can buy 5 cans of green beans at a supermarket for $5.75. How much does it cost to buy 45 cans of green beans? _____

7) The ratio of boys to girls in a class is 4: 7. If there are 32 boys in the class, how many girls are in that class? _____

8) The ratio of red marbles to blue marbles in a bag is 3: 7. If there are 50 marbles in the bag, how many of the marbles are red? _____

Common Core Subject Test Mathematics Grade 8

Percentage Calculations

✎ **Calculate the given percent of each value.**

1) 3% of 60 = ___
2) 20% of 32 = ___
3) 4% of 72 = ___
4) 16% of 32 = ___
5) 25% of 124 = ___
6) 35% of 56 = ___
7) 15% of 20 = ___
8) 14% of 150 = ___
9) 80% of 50 = ___
10) 12% of 115 = ___
11) 72% of 250 = ___
12) 52% of 500 = ___
13) 70% of 400 = ___
14) 27% of 145 = ___
15) 90% of 64 = ___
16) 60% of 55 = ___
17) 22% of 210 = ___
18) 8% of 235 = ___

✎ **Calculate the percent of each given value.**

19) ___% of 25 = 5
20) ___% of 40 = 20
21) ___% of 25 = 2
22) ___% of 50 = 16
23) ___% of 250 = 5
24) ___% of 40 = 32
25) ___% of 125 = 20
26) ___% of 700 = 49
27) ___% of 350 = 49
28) ___% of 500 = 210

✎ **Calculate each percent problem.**

29) A Cinema has 250 seats. 60 seats were sold for the current movie. What percent of seats are empty? ____ %

30) There are 68 boys and 92 girls in a class. 75% of the students in the class take the bus to school. How many students do not take the bus to school? ____

WWW.MathNotion.Com

Common Core Subject Test Mathematics Grade 8

Percent Problems

✎ **Calculate each problem.**

1) 9 is what percent of 45? ___%

2) 60 is what percent of 120? ___%

3) 10 is what percent of 200? ___%

4) 15 is what percent of 125? ___%

5) 10 is what percent of 400? ___%

6) 66 is what percent of 55? ___%

7) 40 is what percent of 160? ___%

8) 40 is what percent of 50? ___%

9) 120 is what percent of 800? ___%

10) 78 is what percent of 120? ___%

11) 36 is what percent of 144? ___%

12) 17 is what percent of 85? ___%

13) 90 is what percent of 900? ___%

14) 36 is what percent of 16? ___%

15) 63 is what percent of 14? ___%

16) 18 is what percent of 60? ___%

17) 126 is what percent of 200? ___%

18) 232 is what percent of 40? ___%

✎ **Calculate each percent word problem.**

19) There are 40 employees in a company. On a certain day, 25 were present. What percent showed up for work? ____%

20) A metal bar weighs 60 ounces. 25% of the bar is gold. How many ounces of gold are in the bar? _____

21) A crew is made up of 12 women; the rest are men. If 15% of the crew are women, how many people are in the crew? _____

22) There are 40 students in a class and 8 of them are girls. What percent are boys? ____%

23) The Royals softball team played 400 games and won 280 of them. What percent of the games did they lose? ____%

WWW.MathNotion.Com

Common Core Subject Test Mathematics Grade 8

Discount, Tax and Tip

✍ **Find the selling price of each item.**

1) Original price of a computer: $420
   Tax: 8% Selling price: $_____

2) Original price of a laptop: $280
   Tax: 4% Selling price: $_____

3) Original price of a sofa: $820
   Tax: 5% Selling price: $_____

4) Original price of a car: $15,800
   Tax: 3.6%   Selling price: $_____

5) Original price of a Table: $250
   Tax: 9% Selling price: $_____

6) Original price of a house: $630,000
   Tax: 1.8%   Selling price: $_____

7) Original price of a tablet: $450
   Discount: 30%   Selling price: $____

8) Original price of a chair: $390
   Discount: 8%   Selling price: $____

9) Original price of a book: $75
   Discount: 42%   Selling price: $____

10) Original price of a cellphone: $820
    Discount: 23%   Selling price: $___

11) Food bill: $45
    Tip: 15%   Price: $_____

12) Food bill: $32
    Tipp: 20%   Price: $_____

13) Food bill: $90
    Tip: 35%   Price: $_____

14) Food bill: $42
    Tipp: 12%   Price: $_____

✍ **Find the answer for each word problem.**

15) Nicolas hired a moving company. The company charged $500 for its services, and Nicolas gives the movers a 40% tip. How much does Nicolas tip the movers? $_____

16) Mason has lunch at a restaurant and the cost of his meal is $90. Mason wants to leave a 25% tip. What is Mason's total bill including tip? $_____

17) The sales tax in Texas is 19.80% and an item costs $350. How much is the tax? $_____

18) The price of a table at Best Buy is $680. If the sales tax is 5%, what is the final price of the table including tax? $_____

WWW.MathNotion.Com

Common Core Subject Test Mathematics Grade 8

Percent of Change

✎ **Find each percent of change.**

1) From 150 to 450. ___ %

2) From 50 ft to 250 ft. ___ %

3) From $60 to $360. ___ %

4) From 60 cm to 180 cm. ___ %

5) From 15 to 45. ___ %

6) From 80 to 16. ___ %

7) From 120 to 360. ___ %

8) From 900 to 450. ___ %

9) From 1,000 to 200. ___ %

10) From 144 to 36. ___ %

✎ **Calculate each percent of change word problem.**

11) Bob got a raise, and his hourly wage increased from $42 to $63. What is the percent increase? ___ %

12) The price of a pair of shoes increases from $50 to $61. What is the percent increase? ___ %

13) At a coffee shop, the price of a cup of coffee increased from $4.80 to $5.76. What is the percent increase in the cost of the coffee? ___ %

14) 51 cm are cut from 85 cm board. What is the percent decrease in length? ___ %

15) In a class, the number of students has been increased from 54 to 81. What is the percent increase? ___ %

16) The price of gasoline rises from $24.40 to $30.50 in one month. By what percent did the gas price rise? ___ %

17) A shirt was originally priced at $38. It went on sale for $24.70. What was the percent that the shirt was discounted? ___ %

Common Core Subject Test Mathematics Grade 8

Simple Interest

🖎 **Determine the simple interest for these loans.**

1) $480 at 11% for 3 years. $ _____

2) $4,200 at 7% for 4 years. $ _____

3) $2,500 at 20% for 3 years. $ _____

4) $6,800 at 3.9% for 4 months. $ _____

5) $800 at 6% for 7 months. $ _____

6) $36,000 at 4.2% for 6 years. $ _____

7) $6,500 at 7% for 4 years. $ _____

8) $850 at 9.5% for 2 years. $ _____

9) $1,200 at 5.8% for 9 months. $ ____

10) $3,000 at 4.5% for 7 years. $ _____

🖎 **Calculate each simple interest word problem.**

11) A new car, valued at $22,000, depreciates at 8.5% per year. What is the value of the car one year after purchase? $_____

12) Sara puts $9,000 into an investment yielding 6% annual simple interest; she left the money in for three years. How much interest does Sara get at the end of those three years? $_____

13) A bank is offering 12% simple interest on a savings account. If you deposit $16,400, how much interest will you earn in two years? $_____

14) $720 interest is earned on a principal of $6,000 at a simple interest rate of 4% interest per year. For how many years was the principal invested? _____

15) In how many years will $2,200 yield an interest of $440 at 4% simple interest? _____

16) Jim invested $8,000 in a bond at a yearly rate of 4.5%. He earned $1,440 in interest. How long was the money invested? _____

WWW.MathNotion.Com

Common Core Subject Test Mathematics Grade 8

Answers of Worksheets

Simplifying Ratios

1) 3 : 4
2) 1 : 10
3) 4 : 7
4) 1 : 3
5) 1 : 10
6) 1 : 8
7) 2 : 8
8) 2 : 5
9) 1 : 6
10) 7 : 9
11) 2 : 3
12) 7 : 2
13) 10 : 1
14) 3 : 4
15) 1 : 8
16) 5 : 7
17) 7 : 9
18) 2 : 1
19) 1 : 3
20) 7 : 1
21) 1 : 2
22) 5 : 4
23) 1 : 2
24) 1 : 30
25) $\frac{1}{2}$
26) $\frac{3}{5}$
27) $\frac{3}{7}$
28) $\frac{1}{3}$
29) $\frac{1}{3}$
30) $\frac{3}{14}$
31) $\frac{1}{2}$
32) $\frac{1}{5}$
33) $\frac{5}{12}$
34) $\frac{2}{9}$
35) $\frac{11}{13}$
36) $\frac{4}{9}$
37) $\frac{1}{12}$
38) $\frac{3}{11}$
39) $\frac{1}{15}$
40) $\frac{1}{4}$
41) $\frac{4}{3}$
42) $\frac{1}{5}$
43) $\frac{22}{41}$
44) $\frac{1}{9}$
45) $\frac{3}{5}$

Proportional Ratios

1) 18
2) 50
3) 9
4) 14
5) 14
6) 56
7) 100
8) 25
9) 70
10) 72
11) 200
12) 77
13) Yes
14) Yes
15) Yes
16) No
17) No
18) No
19) Yes
20) Yes
21) No
22) No
23) Yes
24) Yes
25) 40
26) 256
27) 7
28) 42
29) 63
30) 65
31) 20
32) 104
33) 55
34) 64
35) 126
36) 126

Similarity and ratios

1) 15
2) 5
3) 15
4) 13
5) 150 feet
6) 12.5 meters
7) 50 feet
8) 307.2 miles

Ratio and Rates Word Problems

1) 2 : 3
2) 315

WWW.MathNotion.Com

Common Core Subject Test Mathematics Grade 8

3) The ratio for both classes is 7 to 4.
4) Walmart is a better buy.
5) 720, the rate is 30 per hour.
6) $51.75
7) 56
8) 15

Percentage Calculations

1) 1.8
2) 6.4
3) 2.88
4) 5.12
5) 31
6) 19.6
7) 3
8) 21
9) 40
10) 13.8
11) 180
12) 260
13) 280
14) 39.15
15) 57.6
16) 33
17) 46.2
18) 18.8
19) 20%
20) 50%
21) 8%
22) 32%
23) 2%
24) 80%
25) 16%
26) 7%
27) 14%
28) 42%
29) 76%
30) 40

Percent Problems

1) 20%
2) 50%
3) 5%
4) 12%
5) 2.5%
6) 120%
7) 25%
8) 80%
9) 15%
10) 65%
11) 25%
12) 20%
13) 10%
14) 225%
15) 450%
16) 30%
17) 63%
18) 580%
19) 62.5%
20) 15 ounces
21) 80
22) 80%
23) 30%

Discount, Tax and Tip

1) $453.60
2) $291.20
3) $861.00
4) $16,368.80
5) $272.50
6) $641,340
7) $315.00
8) $358.80
9) $43.50
10) $631.40
11) $51.75
12) $38.40
13) $121.50
14) $47.04
15) $200.00
16) $112.50
17) $69.30
18) $714.00

Common Core Subject Test Mathematics Grade 8

Percent of Change

1) 200%   7) 200%   13) 20%
2) 400%   8) 50%   14) 60%
3) 500%   9) 80%   15) 50%
4) 200%   10) 75%   16) 25%
5) 200%   11) 50%   17) 35%
6) 80%   12) 22%

Simple Interest

1) $158.40   7) $1,820.00   13) $3,936.00
2) $1,176.00   8) $161.50   14) 3 years
3) $1,500.00   9) $52.20   15) 5 years
4) $88.40   10) $945.00   16) 4 years
5) $28.00   11) $20,130.00
6) $9,072.00   12) $1,620.00

Common Core Subject Test Mathematics Grade 8

Chapter 4:
Exponents and Radicals Expressions

Topics that you will practice in this chapter:

- ✓ Multiplication Property of Exponents
- ✓ Zero and Negative Exponents
- ✓ Division Property of Exponents
- ✓ Powers of Products and Quotients
- ✓ Negative Exponents and Negative Bases
- ✓ Scientific Notation
- ✓ Square Roots
- ✓ Simplifying Radical Expressions
- ✓ Simplifying Radical Expressions Involving Fractions
- ✓ Multiplying Radical Expressions
- ✓ Adding and Subtracting Radical Expressions

Mathematics is no more computation than typing is literature.

– John Allen Paulos

Common Core Subject Test Mathematics Grade 8

Multiplication Property of Exponents

✎ Simplify and write the answer in exponential form.

1) $4 \times 4^5 =$

2) $8^4 \times 8 =$

3) $7^3 \times 7^3 =$

4) $9^2 \times 9^2 =$

5) $2^2 \times 2^4 \times 2 =$

6) $5 \times 5^3 \times 5^3 =$

7) $4^3 \times 4^2 \times 4 \times 4 =$

8) $5x \times x =$

9) $x^3 \times x^3 =$

10) $x^7 \times x^2 =$

11) $x^4 \times x^3 \times x^2 =$

12) $10x \times 3x =$

13) $4x^3 \times 4x^3 =$

14) $7x^3 \times x =$

15) $3x^2 \times 4x^2 \times x^2 =$

16) $5x^4 \times x^4 =$

17) $2x^8 \times 2x =$

18) $6x \times x^5 =$

19) $4x^2 \times 6x^6 =$

20) $5yx^3 \times 4x =$

21) $7x^3 \times y^5 x^7 =$

22) $y^2 x^3 \times y^5 x^4 =$

23) $3x^5 \times 4x^3 y^4 =$

24) $4x^4 \times 9x^2 y^5 =$

25) $5x^3 y^4 \times 6x^8 y^2 =$

26) $8x^3 y^6 \times 4xy^3 =$

27) $2xy^5 \times 6x^3 y^3 =$

28) $4x^5 y^2 \times 4x^2 y^8 =$

29) $7x \times 3y^8 x^2 \times y^5 =$

30) $x^3 \times 2y^3 x^4 \times 2y =$

31) $3yx^4 \times 3y^4 x \times 3xy^3 =$

32) $6y^3 \times 2y^2 x^4 \times 10yx^5 =$

WWW.MathNotion.Com

Common Core Subject Test Mathematics Grade 8

Zero and Negative Exponents

✎ Evaluate the following expressions.

1) $1^{-5} =$

2) $4^{-1} =$

3) $0^{10} =$

4) $1^{15} =$

5) $5^{-2} =$

6) $3^{-3} =$

7) $9^{-1} =$

8) $10^{-2} =$

9) $12^{-2} =$

10) $2^{-5} =$

11) $3^{-4} =$

12) $2^{-4} =$

13) $6^{-3} =$

14) $10^{-3} =$

15) $30^{-1} =$

16) $15^{-2} =$

17) $4^{-3} =$

18) $2^{-7} =$

19) $5^{-3} =$

20) $4^{-4} =$

21) $3^{-5} =$

22) $10^{-4} =$

23) $2^{-10} =$

24) $8^{-3} =$

25) $20^{-2} =$

26) $14^{-2} =$

27) $9^{-3} =$

28) $100^{-2} =$

29) $5^{-4} =$

30) $4^{-6} =$

31) $\left(\frac{1}{4}\right)^{-3} =$

32) $\left(\frac{1}{6}\right)^{-2} =$

33) $\left(\frac{1}{7}\right)^{-2} =$

34) $\left(\frac{2}{3}\right)^{-3} =$

35) $\left(\frac{1}{13}\right)^{-2} =$

36) $\left(\frac{7}{12}\right)^{-2} =$

37) $\left(\frac{1}{6}\right)^{-3} =$

38) $\left(\frac{1}{300}\right)^{-2} =$

39) $\left(\frac{2}{9}\right)^{-2} =$

40) $\left(\frac{7}{5}\right)^{-1} =$

41) $\left(\frac{13}{23}\right)^{0} =$

42) $\left(\frac{1}{4}\right)^{-5} =$

WWW.MathNotion.Com

Common Core Subject Test Mathematics Grade 8

Division Property of Exponents

✍ **Simplify.**

1) $\dfrac{5^6}{5^7} =$

2) $\dfrac{8^8}{8^6} =$

3) $\dfrac{4^5}{4} =$

4) $\dfrac{3}{3^5} =$

5) $\dfrac{x}{x^6} =$

6) $\dfrac{3 \times 3^2}{3^2 \times 3^5} =$

7) $\dfrac{9^4}{9^2} =$

8) $\dfrac{10 \times 10^9}{10^2 \times 10^7} =$

9) $\dfrac{7^5 \times 7^7}{7^4 \times 7^8} =$

10) $\dfrac{15x}{30x^6} =$

11) $\dfrac{3x^9}{4x^4} =$

12) $\dfrac{15x^8}{10x^9} =$

13) $\dfrac{42x^5}{6y^9} =$

14) $\dfrac{36y^8}{4x^4y^5} =$

15) $\dfrac{2x^7}{9x} =$

16) $\dfrac{49x^8y^6}{7x^9} =$

17) $\dfrac{48x^2}{24x^6y^{12}} =$

18) $\dfrac{30yx^5}{6yx^7} =$

19) $\dfrac{19x^7y}{38x^{12}y^4} =$

20) $\dfrac{9x^8}{63x^8} =$

21) $\dfrac{9x^{-9}}{4x^{-3}} =$

Powers of Products and Quotients

✏️ **Simplify.**

1) $(4^3)^2 =$

2) $(2^3)^4 =$

3) $(2 \times 2^3)^2 =$

4) $(5 \times 5^5)^6 =$

5) $(19^4 \times 19^2)^3 =$

6) $(2^3 \times 2^4)^4 =$

7) $(5 \times 5^2)^2 =$

8) $(4^4)^4 =$

9) $(8x^5)^2 =$

10) $(3x^2y^4)^4 =$

11) $(7x^5y^2)^2 =$

12) $(5x^4y^4)^3 =$

13) $(2x^3y^3)^5 =$

14) $(10x^3y^4)^3 =$

15) $(13y^3y)^2 =$

16) $(5x^6x^4)^2 =$

17) $(6x^7y^6)^3 =$

18) $(12x^5x^7)^2 =$

19) $(2x^4 \times 2x)^4 =$

20) $(2x^4y^3)^5 =$

21) $(15x^7y^2)^2 =$

22) $(8x^3y^5)^3 =$

23) $(3x \times 2y^2)^4 =$

24) $(\frac{4x}{x^5})^2 =$

25) $(\frac{x^4y^5}{x^3y^5})^9 =$

26) $(\frac{36xy}{6x^5})^3 =$

27) $(\frac{x^7}{x^8y^2})^6 =$

28) $(\frac{xy^4}{x^3y^6})^{-3} =$

29) $(\frac{5xy^8}{x^3})^2 =$

30) $(\frac{xy^6}{2xy^3})^{-4} =$

Negative Exponents and Negative Bases

✏️ **Simplify.**

1) $-9^{-1} =$

2) $-9^{-2} =$

3) $-2^{-5} =$

4) $-x^{-7} =$

5) $11x^{-1} =$

6) $-8x^{-3} =$

7) $-12x^{-5} =$

8) $-9x^{-8}y^{-6} =$

9) $32x^{-5}y^{-1} =$

10) $10a^{-9}b^{-3} =$

11) $-17x^4y^{-6} =$

12) $-\dfrac{25}{x^{-5}} =$

13) $-\dfrac{13x}{a^{-7}} =$

14) $\left(-\dfrac{1}{3}\right)^{-4} =$

15) $\left(-\dfrac{3}{4}\right)^{-2} =$

16) $-\dfrac{14}{a^{-6}b^{-3}} =$

17) $-\dfrac{7x}{x^{-8}} =$

18) $-\dfrac{a^{-9}}{b^{-5}} =$

19) $-\dfrac{11}{x^{-5}} =$

20) $\dfrac{8b}{-16c^{-6}} =$

21) $\dfrac{12ab}{a^{-4}b^{-3}} =$

22) $-\dfrac{8n^{-4}}{32p^{-7}} =$

23) $\dfrac{16ab^{-6}}{-6c^{-5}} =$

24) $\left(\dfrac{10a}{5c}\right)^{-4} =$

25) $\left(-\dfrac{12x}{4yz}\right)^{-3} =$

26) $\dfrac{8ab^{-7}}{-5c^{-3}} =$

27) $\left(-\dfrac{x^4}{x^5}\right)^{-5} =$

28) $\left(-\dfrac{x^{-2}}{7x^3}\right)^{-2} =$

29) $\left(-\dfrac{x^{-4}}{x^2}\right)^{-6} =$

Common Core Subject Test Mathematics Grade 8

Scientific Notation

✎ Write each number in scientific notation.

1) $0.223 =$

2) $0.09 =$

3) $4.5 =$

4) $900 =$

5) $2,000 =$

6) $0.006 =$

7) $33 =$

8) $9,400 =$

9) $1,470 =$

10) $52,000 =$

11) $8,000,000 =$

12) $0.00009 =$

13) $2,158,000 =$

14) $0.0039 =$

15) $0.000075 =$

16) $4,300,000 =$

17) $130,000 =$

18) $4,000,000,000 =$

19) $0.00009 =$

20) $0.0039 =$

✎ Write each number in standard notation.

21) $4 \times 10^{-1} =$

22) $1.2 \times 10^{-3} =$

23) $2.7 \times 10^5 =$

24) $6 \times 10^{-4} =$

25) $3.6 \times 10^{-3} =$

26) $5.5 \times 10^5 =$

27) $3.2 \times 10^4 =$

28) $3.88 \times 10^6 =$

29) $7 \times 10^{-6} =$

30) $4.2 \times 10^{-7} =$

Common Core Subject Test Mathematics Grade 8

Square Roots

✎ **Find the value each square root.**

1) $\sqrt{16} =$ ___

2) $\sqrt{25} =$ ___

3) $\sqrt{1} =$ ___

4) $\sqrt{64} =$ ___

5) $\sqrt{0} =$ ___

6) $\sqrt{196} =$ ___

7) $\sqrt{4} =$ ___

8) $\sqrt{256} =$ ___

9) $\sqrt{36} =$ ___

10) $\sqrt{289} =$ ___

11) $\sqrt{169} =$ ___

12) $\sqrt{144} =$ ___

13) $\sqrt{100} =$ ___

14) $\sqrt{1,600} =$ ___

15) $\sqrt{2,500} =$ ___

16) $\sqrt{324} =$ ___

17) $\sqrt{529} =$ ___

18) $\sqrt{20} =$ ___

19) $\sqrt{625} =$ ___

20) $\sqrt{18} =$ ___

21) $\sqrt{50} =$ ___

22) $\sqrt{1,024} =$ ___

23) $\sqrt{160} =$ ___

24) $\sqrt{32} =$ ___

✎ **Evaluate.**

25) $\sqrt{4} \times \sqrt{25} =$ _____

26) $\sqrt{36} \times \sqrt{49} =$ _____

27) $\sqrt{6} \times \sqrt{6} =$ _____

28) $\sqrt{13} \times \sqrt{13} =$ _____

29) $2\sqrt{5} \times 3\sqrt{5} =$ _____

30) $\sqrt{12} \times \sqrt{3} =$ _____

31) $\sqrt{13} + \sqrt{13} =$ _____

32) $\sqrt{10} + 2\sqrt{10} =$ _____

33) $12\sqrt{7} - 10\sqrt{7} =$ _____

34) $4\sqrt{10} \times 2\sqrt{10} =$ _____

35) $5\sqrt{3} \times 8\sqrt{3} =$ _____

36) $6\sqrt{3} - \sqrt{12} =$ _____

WWW.MathNotion.Com

Simplifying Radical Expressions

✍ **Simplify.**

1) $\sqrt{13x^2} =$

2) $\sqrt{75x^2} =$

3) $\sqrt[3]{27a} =$

4) $\sqrt{64x^5} =$

5) $\sqrt{216a} =$

6) $\sqrt[3]{63w^3} =$

7) $\sqrt{192x} =$

8) $\sqrt{125v} =$

9) $\sqrt[3]{128x^2} =$

10) $\sqrt{100x^9} =$

11) $\sqrt{16x^4} =$

12) $\sqrt[3]{500a^5} =$

13) $\sqrt{242} =$

14) $\sqrt{392p^3} =$

15) $\sqrt{8m^6} =$

16) $\sqrt{198x^3y^3} =$

17) $\sqrt{121x^5y^5} =$

18) $\sqrt{16a^6b^3} =$

19) $\sqrt{90x^5y^7} =$

20) $\sqrt[3]{64y^2x^6} =$

21) $10\sqrt{16x^4} =$

22) $6\sqrt{81x^2} =$

23) $\sqrt[3]{56x^2y^6} =$

24) $\sqrt[3]{1,000x^5y^7} =$

25) $8\sqrt{50a} =$

26) $\sqrt[4]{625x^8y} =$

27) $\sqrt{24x^4y^5r^3} =$

28) $5\sqrt{36x^4y^5z^8} =$

29) $3\sqrt[3]{343x^9y^7} =$

30) $5\sqrt{81a^5b^2c^9} =$

31) $\sqrt[4]{625x^8y^{16}} =$

Common Core Subject Test Mathematics Grade 8

Multiplying Radical Expressions

✎ **Simplify.**

1) $\sqrt{5} \times \sqrt{5} =$

2) $\sqrt{5} \times \sqrt{10} =$

3) $\sqrt{3} \times \sqrt{12} =$

4) $\sqrt{49} \times \sqrt{47} =$

5) $\sqrt{7} \times -2\sqrt{28} =$

6) $3\sqrt{15} \times \sqrt{5} =$

7) $4\sqrt{72} \times \sqrt{2} =$

8) $\sqrt{5} \times -\sqrt{49} =$

9) $\sqrt{55} \times \sqrt{11} =$

10) $7\sqrt{42} \times 2\sqrt{216} =$

11) $\sqrt{45}(5 + \sqrt{5}) =$

12) $\sqrt{13x^2} \times \sqrt{13x^3} =$

13) $-2\sqrt{27} \times \sqrt{3} =$

14) $2\sqrt{13x^4} \times \sqrt{13x^4} =$

15) $\sqrt{14x^3} \times \sqrt{7x^2} =$

16) $-8\sqrt{5x} \times \sqrt{7x^5} =$

17) $-2\sqrt{16x^5} \times 4\sqrt{8x^3} =$

18) $-4\sqrt{32}(8 + \sqrt{32}) =$

19) $\sqrt{32x}(10 - \sqrt{2x}) =$

20) $\sqrt{2x}(8\sqrt{x^5} + \sqrt{8}) =$

21) $\sqrt{20r}(5 + \sqrt{5}) =$

22) $-4\sqrt{7x} \times 3\sqrt{14x^5} =$

23) $-2\sqrt{12x} \times 3\sqrt{2x}$

24) $-\sqrt{7v^3}(-3\sqrt{42v}) =$

25) $(\sqrt{11} - 5)(\sqrt{11} + 5) =$

26) $(-3\sqrt{5} + 3)(\sqrt{5} - 4) =$

27) $(4 - 6\sqrt{3})(-6 + \sqrt{3}) =$

28) $(8 - 3\sqrt{5})(7 - \sqrt{5}) =$

29) $(-1 - \sqrt{3x})(4 + \sqrt{3x}) =$

30) $(-5 + 2\sqrt{7r})(-5 + \sqrt{7r}) =$

31) $(-\sqrt{7n} + 1)(-\sqrt{7} - 5) =$

32) $(-3 + \sqrt{3})(5 - 2\sqrt{3x}) =$

Common Core Subject Test Mathematics Grade 8

Simplifying Radical Expressions Involving Fractions

✎ **Simplify.**

1) $\dfrac{\sqrt{5}}{\sqrt{3}} =$

2) $\dfrac{\sqrt{18}}{\sqrt{45}} =$

3) $\dfrac{\sqrt{10}}{5\sqrt{2}} =$

4) $\dfrac{13}{\sqrt{3}} =$

5) $\dfrac{12\sqrt{5r}}{\sqrt{m^5}} =$

6) $\dfrac{11\sqrt{2}}{\sqrt{k}} =$

7) $\dfrac{6\sqrt{20x^3}}{\sqrt{16x}} =$

8) $\dfrac{\sqrt{14x^3y^4}}{\sqrt{7x^4y^3}} =$

9) $\dfrac{1}{1-\sqrt{5}} =$

10) $\dfrac{1-8\sqrt{a}}{\sqrt{11a}} =$

11) $\dfrac{\sqrt{a}}{\sqrt{a}+\sqrt{b}} =$

12) $\dfrac{1-\sqrt{5}}{2-\sqrt{6}} =$

13) $\dfrac{4+\sqrt{7}}{3-\sqrt{8}} =$

14) $\dfrac{5}{-3-3\sqrt{3}} =$

15) $\dfrac{7}{2-\sqrt{5}} =$

16) $\dfrac{\sqrt{7}-\sqrt{3}}{\sqrt{3}-\sqrt{7}} =$

17) $\dfrac{\sqrt{5}+\sqrt{7}}{\sqrt{7}-\sqrt{5}} =$

18) $\dfrac{2\sqrt{2}-\sqrt{3}}{3\sqrt{2}+\sqrt{5}} =$

19) $\dfrac{\sqrt{11}+5\sqrt{3}}{4-\sqrt{11}} =$

20) $\dfrac{\sqrt{5}+\sqrt{3}}{2-\sqrt{3}} =$

21) $\dfrac{\sqrt{32a^7b^4}}{\sqrt{2ab^3}} =$

22) $\dfrac{10\sqrt{21x^5}}{5\sqrt{x^3}} =$

Common Core Subject Test Mathematics Grade 8

Adding and Subtracting Radical Expressions

✏️ **Simplify.**

1) $\sqrt{2} + \sqrt{8} =$

2) $3\sqrt{50} + 4\sqrt{2} =$

3) $2\sqrt{12} - 4\sqrt{3} =$

4) $5\sqrt{32} - 5\sqrt{2} =$

5) $3\sqrt{75} - 5\sqrt{3} =$

6) $-\sqrt{72} - 4\sqrt{2} =$

7) $-7\sqrt{16} - 4\sqrt{25} =$

8) $8\sqrt{24} + 2\sqrt{6} =$

9) $10\sqrt{49} - 7\sqrt{100} =$

10) $-7\sqrt{5} + 9\sqrt{45} =$

11) $-15\sqrt{12} + 14\sqrt{48} =$

12) $20\sqrt{4} - 2\sqrt{25} =$

13) $-2\sqrt{20} + 7\sqrt{5} =$

14) $8\sqrt{7} - 2\sqrt{63} =$

15) $5\sqrt{44} + 3\sqrt{11} =$

16) $3\sqrt{27} - 5\sqrt{48} =$

17) $\sqrt{144} - \sqrt{81} =$

18) $3\sqrt{20} - 6\sqrt{5} =$

19) $-2\sqrt{7} + 8\sqrt{28} =$

20) $3\sqrt{75} - 2\sqrt{3} =$

21) $5\sqrt{27} - 3\sqrt{3} =$

22) $-7\sqrt{30} + 6\sqrt{120} =$

23) $-7\sqrt{24} - 2\sqrt{6} =$

24) $-\sqrt{32x} + 4\sqrt{2x} =$

25) $\sqrt{7y^2} + y\sqrt{112} =$

26) $\sqrt{45mn^2} + 2n\sqrt{5m} =$

27) $-4\sqrt{12a} - 4\sqrt{3a} =$

28) $-5\sqrt{15ab} - 2\sqrt{60ab} =$

29) $\sqrt{45x^2y} + x\sqrt{20y} =$

30) $2\sqrt{7a} + 4\sqrt{63a} =$

WWW.MathNotion.Com

Common Core Subject Test Mathematics Grade 8

Answers of Worksheets

Multiplication Property of Exponents

1) 4^6
2) 8^5
3) 7^6
4) 9^4
5) 2^7
6) 5^7
7) 4^7
8) $5x^2$
9) x^6
10) x^9
11) x^9
12) $30x^2$
13) $16x^6$
14) $7x^4$
15) $12x^6$
16) $5x^8$
17) $4x^9$
18) $6x^6$
19) $24x^8$
20) $20x^4y$
21) $7x^{10}y^5$
22) x^7y^7
23) $12x^8y^4$
24) $36x^6y^5$
25) $30x^{11}y^6$
26) $32x^4y^9$
27) $12x^4y^8$
28) $16x^7y^{10}$
29) $21x^3y^{13}$
30) $4x^7y^4$
31) $27x^6y^8$
32) $120x^9y^6$

Zero and Negative Exponents

1) 1
2) $\frac{1}{4}$
3) 0
4) 1
5) $\frac{1}{25}$
6) $\frac{1}{27}$
7) $\frac{1}{9}$
8) $\frac{1}{100}$
9) $\frac{1}{144}$
10) $\frac{1}{32}$
11) $\frac{1}{81}$
12) $\frac{1}{16}$
13) $\frac{1}{216}$
14) $\frac{1}{1,000}$
15) $\frac{1}{30}$
16) $\frac{1}{225}$
17) $\frac{1}{64}$
18) $\frac{1}{128}$
19) $\frac{1}{125}$
20) $\frac{1}{256}$
21) $\frac{1}{243}$
22) $\frac{1}{10,000}$
23) $\frac{1}{1,024}$
24) $\frac{1}{512}$
25) $\frac{1}{400}$
26) $\frac{1}{196}$
27) $\frac{1}{729}$
28) $\frac{1}{10,000}$
29) $\frac{1}{625}$
30) $\frac{1}{4,096}$
31) 64
32) 36
33) 49
34) $\frac{27}{8}$
35) 169
36) $\frac{144}{49}$
37) 216
38) $90,000$
39) $\frac{81}{4}$
40) $\frac{5}{7}$
41) 1
42) $1,024$

Division Property of Exponents

1) $\frac{1}{5}$
2) 8^2
3) 4^4
4) $\frac{1}{3^4}$
5) $\frac{1}{x^5}$
6) $\frac{1}{3^4}$
7) 9^2
8) 10
9) 1
10) $\frac{1}{2x^5}$
11) $\frac{3x^5}{4}$
12) $\frac{3}{2x}$
13) $\frac{7x^5}{y^9}$
14) $\frac{9y^3}{x^4}$
15) $\frac{2x^6}{9}$

WWW.MathNotion.Com

Common Core Subject Test Mathematics Grade 8

16) $\dfrac{7y^6}{x}$ 18) $\dfrac{5}{x^2}$ 19) $\dfrac{1}{2x^5y^3}$ 21) $\dfrac{9}{4x^6}$

17) $\dfrac{2}{x^4y^{12}}$ 20) $\dfrac{1}{7}$

Powers of Products and Quotients

1) 4^6
2) 2^{12}
3) 2^8
4) 5^{36}
5) 19^{18}
6) 2^{28}
7) 5^6
8) 4^{16}
9) $64x^{10}$
10) $81x^8y^{16}$
11) $49x^{10}y^4$
12) $125x^{12}y^{12}$
13) $32x^{15}y^{15}$
14) $1{,}000x^9y^{12}$
15) $169y^8$
16) $25x^{20}$
17) $216x^{21}y^{18}$
18) $144x^{24}$
19) $256x^{20}$
20) $32x^{20}y^{15}$
21) $225x^{14}y^4$
22) $512x^9y^{15}$
23) $1{,}296x^4y^8$
24) $\dfrac{16}{x^8}$
25) x^9
26) $\dfrac{216y^3}{x^{12}}$
27) $\dfrac{1}{x^6y^{12}}$
28) x^6y^6
29) $\dfrac{25y^{16}}{x^4}$
30) $\dfrac{16}{y^{12}}$

Negative Exponents and Negative Bases

1) $-\dfrac{1}{9}$
2) $-\dfrac{1}{81}$
3) $-\dfrac{1}{32}$
4) $-\dfrac{1}{x^7}$
5) $\dfrac{11}{x}$
6) $-\dfrac{8}{x^3}$
7) $-\dfrac{12}{x^5}$
8) $-\dfrac{9}{x^8y^6}$
9) $\dfrac{32}{x^5y}$
10) $\dfrac{10}{a^9b^3}$
11) $-\dfrac{17x^4}{y^6}$
12) $-25x^5$
13) $-13xa^7$
14) 81
15) $\dfrac{16}{9}$
16) $-14a^6b^3$
17) $-7x^9$
18) $-\dfrac{b^5}{a^9}$
19) $-11x^5$
20) $-\dfrac{bc^6}{2}$
21) $12a^5b^4$
22) $-\dfrac{p^7}{4n^4}$
23) $-\dfrac{8ac^5}{3b^6}$
24) $\dfrac{c^4}{16a^4}$
25) $\dfrac{y^3z^3}{27x^3}$
26) $-\dfrac{8ac^3}{5b^7}$
27) $-x^5$
28) $49x^{10}$
29) x^{36}

Scientific Notation

Common Core Subject Test Mathematics Grade 8

1) 2.23×10^{-1}	11) 8×10^6	21) 0.4
2) 9×10^{-2}	12) 9×10^{-5}	22) 0.0012
3) 4.5×10^0	13) 2.158×10^6	23) 270,000
4) 9×10^2	14) 3.9×10^{-3}	24) 0.0006
5) 2×10^3	15) 7.5×10^{-5}	25) 0.0036
6) 6×10^{-3}	16) 4.3×10^6	26) 550,000
7) 3.3×10^1	17) 1.3×10^5	27) 32,000
8) 9.4×10^3	18) 4×10^9	28) 3,880,000
9) 1.47×10^3	19) 9×10^{-5}	29) 0.000007
10) 5.2×10^4	20) 3.9×10^{-3}	30) 0.00000042

Square Roots

1) 4	10) 17	19) 25	28) 13
2) 5	11) 13	20) $3\sqrt{2}$	29) 30
3) 1	12) 12	21) $5\sqrt{2}$	30) 6
4) 8	13) 10	22) 32	31) $2\sqrt{13}$
5) 0	14) 40	23) $4\sqrt{10}$	32) $3\sqrt{10}$
6) 14	15) 50	24) $4\sqrt{2}$	33) $2\sqrt{7}$
7) 2	16) 18	25) 10	34) 80
8) 16	17) 23	26) 42	35) 120
9) 6	18) $2\sqrt{5}$	27) 6	36) $4\sqrt{3}$

Simplifying radical expressions

1) $x\sqrt{13}$	9) $4\sqrt[3]{2x^2}$	17) $11x^2y^2\sqrt{xy}$
2) $5x\sqrt{3}$	10) $10x^4\sqrt{x}$	18) $4a^3b\sqrt{b}$
3) $3\sqrt[3]{a}$	11) $4x^2$	19) $3x^2y^3\sqrt{10xy}$
4) $8x^2\sqrt{x}$	12) $5a\sqrt[3]{4a^2}$	20) $4x^2\sqrt[3]{y^2}$
5) $6\sqrt{6a}$	13) $11\sqrt{2}$	21) $40x^2$
6) $w\sqrt[3]{63}$	14) $14p\sqrt{2p}$	22) $54x$
7) $8\sqrt{3x}$	15) $2m^3\sqrt{2}$	23) $2y^2\sqrt[3]{7x^2}$
8) $5\sqrt{5v}$	16) $3x.y\sqrt{22xy}$	24) $10xy^2\sqrt[3]{x^2y}$

Common Core Subject Test Mathematics Grade 8

25) $40\sqrt{2a}$

26) $5x^2\sqrt[4]{y}$

27) $2x^2y^2r\sqrt{6yr}$

28) $30x^2y^2z^4\sqrt{y}$

29) $21x^3y^2\sqrt[3]{y}$

30) $45a^2bc^4\sqrt{ac}$

31) $5x^2y^4$

Multiplying radical expressions

1) 5
2) $5\sqrt{2}$
3) 6
4) $7\sqrt{47}$
5) -28
6) $15\sqrt{3}$
7) 48
8) $-5\sqrt{7}$
9) $11\sqrt{5}$
10) $504\sqrt{7}$
11) $15\sqrt{5}+15$
12) $13x^2\sqrt{x}$
13) -18
14) $26x^4$
15) $7x^2\sqrt{2x}$
16) $-8x^3\sqrt{35}$
17) $-64x^4\sqrt{2}$
18) $-128\sqrt{2}-128$
19) $40\sqrt{2x}-8x$
20) $8x^3\sqrt{2}+4\sqrt{x}$
21) $10\sqrt{5r}+10\sqrt{r}$
22) $-84x^3\sqrt{2}$
23) $-12\sqrt{6}x$
24) $21v^2\sqrt{6}$
25) -14
26) $15\sqrt{5}-27$
27) $40\sqrt{3}-42$
28) $71-29\sqrt{5}$
29) $-3x-5\sqrt{3x}-4$
30) $14r-15\sqrt{7r}+25$
31) $7\sqrt{n}+5\sqrt{7n}-\sqrt{7}-5$
32) $-15+6\sqrt{3x}+5\sqrt{3}-6\sqrt{x}$

Simplifying radical expressions involving fractions

1) $\frac{\sqrt{15}}{3}$
2) $\frac{9\sqrt{10}}{45}=\frac{\sqrt{10}}{5}$
3) $\frac{\sqrt{20}}{10}=\frac{\sqrt{5}}{5}$
4) $\frac{13\sqrt{3}}{3}$
5) $\frac{12\sqrt{5mr}}{m^3}$
6) $\frac{11\sqrt{2k}}{k}$
7) $3x\sqrt{5}$
8) $\frac{\sqrt{2x}}{xy}$
9) $\frac{-1-\sqrt{5}}{4}$
10) $\frac{\sqrt{11a}-8a\sqrt{11}}{11a}$

Common Core Subject Test Mathematics Grade 8

11) $\frac{a-\sqrt{ab}}{a-b}$

12) $\frac{\sqrt{30}+2\sqrt{5}-\sqrt{6}-2}{2}$

13) $12 + 8\sqrt{2} + 3\sqrt{7} + 2\sqrt{14}$

14) $-\frac{5(\sqrt{3}-1)}{6}$

15) $-14 - 7\sqrt{5}$

16) -1

17) $6 + \sqrt{35}$

18) $\frac{12 - 2\sqrt{10} - 3\sqrt{6} + \sqrt{15}}{13}$

19) $\frac{4\sqrt{11}+11+20\sqrt{3}+5\sqrt{33}}{5}$

20) $2\sqrt{5} + 3 + \sqrt{15} + 2\sqrt{3}$

21) $4a^3\sqrt{b}$

22) $2x\sqrt{21}$

Adding and subtracting radical expressions

1) $3\sqrt{2}$
2) $19\sqrt{2}$
3) 0
4) $15\sqrt{2}$
5) $10\sqrt{3}$
6) $-10\sqrt{2}$
7) -48
8) $18\sqrt{6}$
9) 0
10) $20\sqrt{5}$

11) $26\sqrt{3}$
12) 30
13) $3\sqrt{5}$
14) $2\sqrt{7}$
15) $13\sqrt{11}$
16) $-11\sqrt{3}$
17) 3
18) 0
19) $14\sqrt{7}$
20) $13\sqrt{3}$

21) $12\sqrt{3}$
22) $5\sqrt{30}$
23) $-16\sqrt{6}$
24) 0
25) $5y\sqrt{7}$
26) $5n\sqrt{5m}$
27) $-12\sqrt{3a}$
28) $-9\sqrt{15ab}$
29) $5x\sqrt{5y}$
30) $14\sqrt{7a}$

WWW.MathNotion.Com

Common Core Subject Test Mathematics Grade 8

Chapter 5 :
Algebraic Expressions

Topics that you will practice in this chapter:

- ✓ Simplifying Variable Expressions
- ✓ Simplifying Polynomial Expressions
- ✓ Translate Phrases into an Algebraic Statement
- ✓ The Distributive Property
- ✓ Evaluating One Variable Expressions
- ✓ Evaluating Two Variables Expressions
- ✓ Combining like Terms

Mathematics is, as it were, a sensuous logic, and relates to philosophy as do the arts, music, and plastic art to poetry. — *K. Shegel*

Common Core Subject Test Mathematics Grade 8

Simplifying Variable Expressions

✎ Simplify each expression.

1) $3(x + 5) =$

2) $(-4)(7x - 5) =$

3) $11x + 5 - 6x =$

4) $-4 - 2x^2 - 6x^2 =$

5) $7 + 13x^2 + 3 =$

6) $3x^2 + 7x + 15x^2 =$

7) $3x^2 - 12x^2 + 4x =$

8) $4x^2 - 8x - 2x =$

9) $6x + 7(3 - 4x) =$

10) $8x + 4(15x - 3) =$

11) $6(-3x - 9) - 17 =$

12) $-11x^2 - (-5x) =$

13) $2x + 7 + 5 - 8x =$

14) $7 + 6x - 11 - 5x =$

15) $27x + 8 - 13 - 5x =$

16) $(-11)(-5x + 2) - 41x =$

17) $19x - 4(4 - 2x) =$

18) $16x + 3(3x + 6) + 10 =$

19) $5(-2x - 4) - 13x =$

20) $16x - 3x(x + 10) =$

21) $17x + 5x(2 - 4x) =$

22) $5x(-4x - 7) + 20x =$

23) $25x - 19 + 4x^2 =$

24) $6x(x - 11) + 25 =$

25) $4x - 5 + 15x + 3x^2 =$

26) $-7x^2 - 11x - 9x =$

27) $10x - 9x^2 - 3x^2 - 7 =$

28) $13 + 3x^2 - 9x^2 - 21x =$

29) $22x + 10x^2 - 15x + 17 =$

30) $4x^2 + 25x + 21x^2 =$

31) $29 - 12x^2 - 23x - 4x^2 =$

32) $22x - 19x - 9x^2 + 30 =$

WWW.MathNotion.Com

Common Core Subject Test Mathematics Grade 8

Simplifying Polynomial Expressions

✎ Simplify each polynomial.

1) $(2x^3 + 8x^2) - (11x + 3x^2) =$ _____

2) $(2x^5 + 7x^3) - (5x^3 + 11x^2) =$ _____

3) $(41x^4 + 5x^2) - (4x^2 + 20x^4) =$ _____

4) $13x - 8x^2 + 4(4x^2 + 3x^3) =$ _____

5) $(4x^3 - 22) + 5(3x^2 - 6x^3) =$ _____

6) $(4x^3 - 3x) - 5(2x^3 + x^4) =$ _____

7) $5(5x - 2x^3) - 2(8x^3 + 5x^2) =$ _____

8) $(3x^2 - 10x) - (5x^3 + 14x^2) =$ _____

9) $5x^3 - (3x^4 + 5x) + 2x^2 =$ _____

10) $11x^4 - (3x^2 + 5x) + 7x =$ _____

11) $(6x^2 - 3x^4) - (10x^4 + 3x^2) =$ _____

12) $2x^2 - 7x^3 + 19x^4 - 22x^3 =$ _____

13) $10x^2 - x^4 + 4x^4 - 32x^3 =$ _____

14) $-5x^2 + 17x^3 - 8x^2 - 6x =$ _____

15) $x^4 - 11x^5 - 30x^4 + 5x^2 =$ _____

16) $21x^3 + 13x - 5x^2 - 11x^3 =$ _____

WWW.MathNotion.Com

Common Core Subject Test Mathematics Grade 8

Translate Phrases into an Algebraic Statement

✏ **Write an algebraic expression for each phrase.**

1) 9 multiplied by x. _____

2) Subtract 11 from y. _____

3) 19 divided by x. _____

4) 38 decreased by y. _____

5) Add y to 40. _____

6) The square of 6. _____

7) x raised to the fifth power. _____

8) The sum of six and a number. _____

9) The difference between fifty-seven and y. _____

10) The quotient of nine and a number. _____

11) The quotient of the square of x and 25. _____

12) The difference between x and 6 is 19. _____

13) 10 times a reduced by the square of b. _____

14) Subtract the product of a and b from 41. _____

The Distributive Property

✏ Use the distributive property to simply each expression.

1) $4(1 + 2x) =$

2) $2(4 + 7x) =$

3) $3(4x - 4) =$

4) $(2x - 5)(-6) =$

5) $(-3)(x + 6) =$

6) $(4 + 3x)2 =$

7) $(-5)(8 - 3x) =$

8) $-(-5 - 7x) =$

9) $(-6x + 3)(-3) =$

10) $(-4)(x - 7) =$

11) $-(5 - 3x) =$

12) $3(9 + 4x) =$

13) $6(4 + 3x) =$

14) $(-5x + 3)2 =$

15) $(5 - 8x)(-3) =$

16) $(-12)(3x + 3) =$

17) $(5 - 3x)6 =$

18) $4(2 + 6x) =$

19) $8(7x - 3) =$

20) $(-2x + 3)4 =$

21) $(7 - 5x)(-9) =$

22) $(-10)(x - 8) =$

23) $(11 - 4x)3 =$

24) $(-6)(10x - 4) =$

25) $(3 - 9x)(-7) =$

26) $(-9)(x + 9) =$

27) $(-3 + 5x)(-7) =$

28) $(-5)(8 - 10x) =$

29) $12(4x - 8) =$

30) $(-10x + 13)(-3) =$

31) $(-8)(3x - 2) + 4(x + 5) =$

32) $(-8)(x + 4) - (6 + 5x) =$

Common Core Subject Test Mathematics Grade 8

Evaluating One Variable Expressions

✎ **Evaluate each expression using the value given.**

1) $8 - x, x = 5$

2) $x - 9, x = 5$

3) $5x + 4, x = 3$

4) $x - 13, x = -4$

5) $12 - x, x = 4$

6) $x + 2, x = 6$

7) $4x + 8, x = 3$

8) $x + (-7), x = -8$

9) $4x + 5, x = 2$

10) $3x + 9, x = -2$

11) $15 + 3x - 7, x = 2$

12) $17 - 3x, x = 3$

13) $8x - 9, x = 4$

14) $5x + 4, x = -3$

15) $10x + 5, x = 3$

16) $14 - 4x, x = -6$

17) $3(5x + 3), x = 9$

18) $4(-3x - 6), x = 3$

19) $7x - 2x + 12, x = 4$

20) $(5x + 6) \div 2, x = 8$

21) $(x + 18) \div 10, x = 12$

22) $5x - 12 + 3x, x = -3$

23) $(6 - 4x)(-3), x = -4$

24) $9x^2 + 3x - 6, x = 2$

25) $x^2 - 10x, x = -5$

26) $3x(7 - 2x), x = 2$

27) $12x + 6 - 2x^2, x = -4$

28) $(-3)(4x - 8 + 3x), x = 3$

29) $(-6) + \frac{x}{4} + 3x, x = 16$

30) $(-6) + \frac{x}{5}, x = 35$

31) $\left(-\frac{45}{x}\right) - 7 + 2x, x = 9$

32) $\left(-\frac{21}{x}\right) - 12 + 4x, x = 7$

Evaluating Two Variables Expressions

✎ Evaluate each expression using the values given.

1) $2x - 4y$,
   $x = 4, y = 1$

2) $3x + 5y$,
   $x = -2, y = 2$

3) $-7a + 4b$,
   $a = 2, b = 4$

4) $3x + 5 - y$,
   $x = 5, y = 6$

5) $3z + 12 - 2k$,
   $z = 5, k = 6$

6) $6(-x - 3y)$,
   $x = 5, y = -2$

7) $5a + 3b$,
   $a = 3, b = 4$

8) $7x \div 3y$,
   $x = 3, y = 7$

9) $2x + 15 + 5y$,
   $x = -3, y = 1$

10) $5a - (18 - b)$,
    $a = 2, b = 8$

11) $2z + 20 + 5k$,
    $z = -6, k = 5$

12) $xy + 10 + 4x$,
    $x = 3, y = 5$

13) $2x + 4y - 8 + 5$,
    $x = 5, y = 2$

14) $\left(-\frac{24}{x}\right) + 3 + 2y$,
    $x = 4, y = 6$

15) $(-3)(-3a - 3b)$,
    $a = 4, b = 5$

16) $12 + 4x - 7 - y$,
    $x = 3, y = 5$

17) $11x + 5 - 8y + 6$,
    $x = 5, y = 2$

18) $10 + 2(-4x - 5y)$,
    $x = 5, y = 4$

19) $5x + 13 + 6y$,
    $x = 5, y = 6$

20) $10a - (7a + 3b) - 11$,
    $a = 3, b = 8$

Common Core Subject Test Mathematics Grade 8

Combining like Terms

✎ **Simplify each expression.**

1) $11x + 3x + 6 =$

2) $8(2x - 6) =$

3) $18x - 7x + 11 =$

4) $(-4)(6x - 7) =$

5) $22x - 10x - 5 =$

6) $32x - 13 + 8x =$

7) $15 - (8x - 11) =$

8) $-24x + 17 - 11x =$

9) $12x - 8 - 6x + 9 =$

10) $21x + 5 - 36 + 12x =$

11) $28x + 3x - 11 =$

12) $(-3x + 4)5 =$

13) $2 + 4x + 9x - 8 =$

14) $6(2x - 5x) - 4 =$

15) $4(5x + 11) + 3x =$

16) $x - 14 - 11x =$

17) $5(10 + 9x) - 8x =$

18) $42x + 17 - 23x =$

19) $(-7x) + 19 + 20x =$

20) $(-7x) - 33 + 29x =$

21) $4(5x + 3) - 19x =$

22) $5(6 - 2x) - 15x =$

23) $-24x + (11 - 18x) =$

24) $(-9) - (6)(7x + 3) =$

25) $(-1)(8x - 10) - 21x =$

26) $-36x + 14 + 27x - 5x =$

27) $3(-13x + 6) - 17x =$

28) $-5x - 42 + 32x =$

29) $37x - 19x + 15 - 9x =$

30) $3(5x + 7x) - 31 =$

31) $14 - 6x - 15 - 9x =$

32) $-2(-5x - 7x) + 27x =$

Common Core Subject Test Mathematics Grade 8

Answers of Worksheets

Simplifying Variable Expressions

1) $3x + 15$
2) $-28x + 20$
3) $5x + 5$
4) $-8x^2 - 4$
5) $13x^2 + 10$
6) $18x^2 + 7x$
7) $-9x^2 + 4x$
8) $4x^2 - 10x$
9) $-22x + 21$
10) $68x - 12$
11) $-18x - 71$
12) $-11x^2 + 5x$
13) $-6x + 12$
14) $x - 4$
15) $22x - 5$
16) $14x - 22$
17) $27x - 16$
18) $25x + 28$
19) $-23x - 20$
20) $-3x^2 - 14x$
21) $-20x^2 + 27x$
22) $-20x^2 - 15x$
23) $4x^2 + 25x - 19$
24) $6x^2 - 66x + 25$
25) $3x^2 + 19x - 5$
26) $-7x^2 - 20x$
27) $-12x^2 + 10x - 7$
28) $-6x^2 - 21x + 13$
29) $10x^2 + 7x + 17$
30) $25x^2 + 25x$
31) $-16x^2 - 23x + 29$
32) $-9x^2 + 3x + 30$

Simplifying Polynomial Expressions

1) $2x^3 + 5x^2 - 11x$
2) $2x^5 + 2x^3 - 11x^2$
3) $21x^4 + x^2$
4) $12x^3 + 8x^2 + 13x$
5) $-26x^3 + 15x^2 - 22$
6) $-5x^4 - 6x^3 - 3x$
7) $-26x^3 - 10x^2 + 25x$
8) $-5x^3 - 11x^2 - 10x$
9) $-3x^4 + 5x^3 + 2x^2 - 5x$
10) $11x^4 - 3x^2 + 2x$
11) $-13x^4 + 3x^2$
12) $19x^4 - 29x^3 + 2x^2$
13) $3x^4 - 32x^3 + 10x^2$
14) $17x^3 - 13x^2 - 6x$
15) $-11x^5 - 29x^4 + 5x^2$
16) $10x^3 - 5x^2 + 13x$

Translate Phrases into an Algebraic Statement

1) $9x$
2) $y - 11$
3) $\frac{19}{x}$
4) $38 - y$
5) $y + 40$
6) 6^2
7) x^5
8) $6 + x$
9) $57 - y$
10) $\frac{9}{x}$
11) $\frac{x^2}{25}$
12) $x - 6 = 19$
13) $10a - b^2$
14) $41 - ab$

The Distributive Property

1) $8x + 4$
2) $14x + 8$
3) $12x - 12$
4) $-12x + 30$
5) $-3x - 18$
6) $6x + 8$
7) $15x - 40$
8) $7x + 5$
9) $18x - 9$
10) $-4x + 28$
11) $3x - 5$
12) $12x + 27$

Common Core Subject Test Mathematics Grade 8

13) $18x + 24$
14) $-10x + 6$
15) $24x - 15$
16) $-36x - 36$
17) $-18x + 30$

18) $24x + 8$
19) $56x - 24$
20) $-8x + 12$
21) $45x - 63$
22) $-10x + 80$

23) $-12x + 33$
24) $-60x + 24$
25) $63x - 21$
26) $-9x - 81$
27) $-35x + 21$

28) $50x - 40$
29) $48x - 96$
30) $30x - 39$
31) $-20x + 36$
32) $-13x - 38$

Evaluating One Variables

1) 3
2) -4
3) 19
4) -17
5) 8
6) 8
7) 20
8) -15

9) 13
10) 3
11) 14
12) 8
13) 23
14) -11
15) 35
16) 38

17) 144
18) -60
19) 32
20) 23
21) 3
22) -36
23) -66
24) 36

25) 75
26) 18
27) -74
28) -39
29) 46
30) 1
31) 6
32) 13

Evaluating Two Variables

1) 4
2) 4
3) 2
4) 14
5) 15

6) 6
7) 27
8) 1
9) 14
10) 0

11) 33
12) 37
13) 15
14) 9
15) 81

16) 12
17) 50
18) -70
19) 74
20) -26

Combining like Terms

1) $14x + 6$
2) $16x - 48$
3) $11x + 11$
4) $-24x + 28$
5) $12x - 5$
6) $40x - 13$
7) $-8x + 26$
8) $-35x + 17$

9) $6x + 1$
10) $33x - 31$
11) $31x - 11$
12) $-15x + 20$
13) $13x - 6$
14) $-18x - 4$
15) $23x + 44$
16) $-10x - 14$

17) $37x + 50$
18) $19x + 17$
19) $13x + 19$
20) $22x - 33$
21) $x + 12$
22) $-25x + 30$
23) $-42x + 11$
24) $-42x - 27$

25) $-29x + 10$
26) $-14x + 14$
27) $-56x + 18$
28) $27x - 42$
29) $9x + 15$
30) $36x - 31$
31) $-15x - 1$
32) $51x$

Common Core Subject Test Mathematics Grade 8

Chapter 6
Equations and Inequalities

Topics that you will practice in this chapter:

- ✓ One–Step Equations
- ✓ Multi–Step Equations
- ✓ Graphing Single–Variable Inequalities
- ✓ One–Step Inequalities
- ✓ Multi-Step Inequalities
- ✓ Systems of Equations
- ✓ Systems of Equations Word Problems

"Life is a math equation. In order to gain the most, you have to know how to convert negatives into positives." – Anonymous

Common Core Subject Test Mathematics Grade 8

One–Step Equations

✏ **Find the answer for each equation.**

1) $3x = 90$, $x =$ ___

2) $5x = 35$, $x =$ ___

3) $6x = 24$, $x =$ ___

4) $24x = 144$, $x =$ ___

5) $x + 15 = 20$, $x =$ ___

6) $x - 7 = 4$, $x =$ ___

7) $x - 9 = 2$, $x =$ ___

8) $x + 15 = 23$, $x =$ ___

9) $x - 4 = 13$, $x =$ ___

10) $12 = 16 + x$, $x =$ ___

11) $x - 10 = 2$, $x =$ ___

12) $5 - x = -11$, $x =$ ___

13) $28 = -6 + x$, $x =$ ___

14) $x - 20 = -35$, $x =$ ___

15) $x + 14 = -4$, $x =$ ___

16) $14 = 28 - x$, $x =$ ___

17) $7 + x = -7$, $x =$ ___

18) $x - 16 = 4$, $x =$ ___

19) $30 = x - 15$, $x =$ ___

20) $x - 5 = -18$, $x =$ ___

21) $x - 10 = 24$, $x =$ ___

22) $x - 20 = -25$, $x =$ ___

23) $x - 17 = 30$, $x =$ ___

24) $-70 = x - 28$, $x =$ ___

25) $x - 9 = 13$, $x =$ ___

26) $36 = 4x$, $x =$ ___

27) $x - 35 = 25$, $x =$ ___

28) $x - 25 = 10$, $x =$ ___

29) $70 - x = 16$, $x =$ ___

30) $x - 10 = 14$, $x =$ ___

31) $17 - x = -13$, $x =$ ___

32) $x - 9 = -30$, $x =$ ___

WWW.MathNotion.Com

Common Core Subject Test Mathematics Grade 8

Multi–Step Equations

✏ **Find the answer for each equation.**

1) $3x + 3 = 9$

2) $-x + 5 = 12$

3) $4x - 8 = 8$

4) $-(3 - x) = 5$

5) $4x - 8 = 16$

6) $12x - 15 = 9$

7) $2x - 18 = 2$

8) $4x + 8 = 16$

9) $24x + 27 = 75$

10) $-14(3 + x) = 14$

11) $-3(2 + x) = 6$

12) $12 = -(x - 7)$

13) $3(3 - x) = 30$

14) $-15 = -(3x + 6)$

15) $40(3 + x) = 40$

16) $5(x - 10) = 25$

17) $-18 = x + 8x$

18) $3x + 25 = -2x - 10$

19) $7(6 + 3x) = -63$

20) $18 - 3x = -4 - 5x$

21) $4 - 6x = 36 + 2x$

22) $15 + 15x = -5 + 5x$

23) $42 = (-6x) - 7 + 7$

24) $21 = 3x - 21 + 4x$

25) $-18 = -6x - 9 + 3x$

26) $5x - 15 = -29 + 6x$

27) $7x - 18 = 4x + 3$

28) $-7 - 4x = 5(4 - x)$

29) $x - 5 = -5(-3 - x)$

30) $13x - 68 = 15x - 102$

31) $-5x - 3 = -3(9 + 3x)$

32) $-2x - 15 = 6x + 17$

Common Core Subject Test Mathematics Grade 8

Graphing Single–Variable Inequalities

✎ **Draw a graph for each inequality.**

1) $x > -1$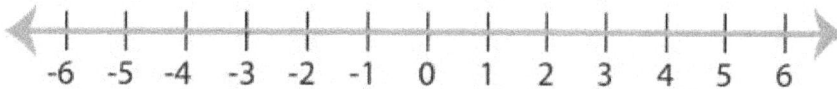

2) $x \leq 2$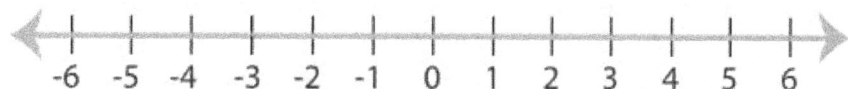

3) $x \geq 0$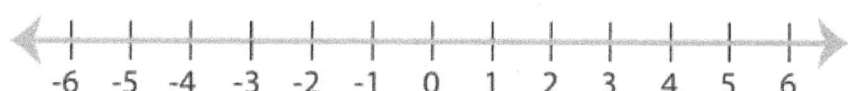

4) $x < -3$

5) $x < \frac{1}{2}$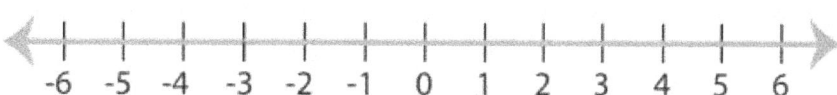

6) $x \leq -2$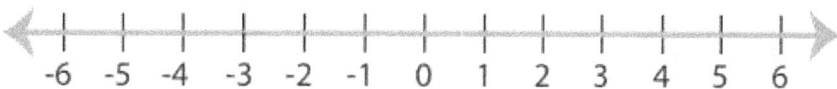

7) $x \leq 3$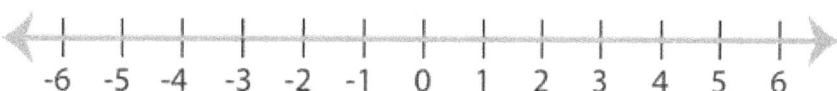

8) $x \geq -\frac{7}{2}$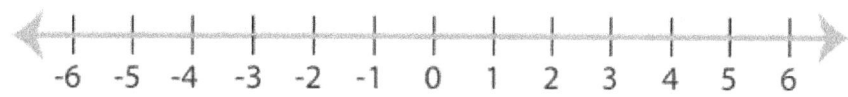

WWW.MathNotion.Com

Common Core Subject Test Mathematics Grade 8

One–Step Inequalities

✎ **Find the answer for each inequality and graph it.**

1) $x + 4 \geq 4$

2) $x - 5 \leq 2$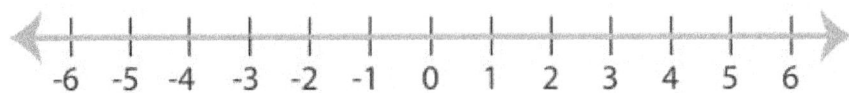

3) $5x > 35$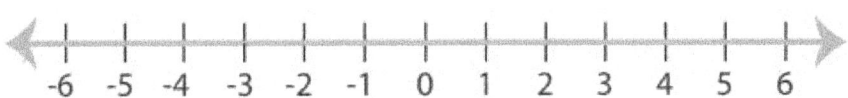

4) $9 + x \leq 11$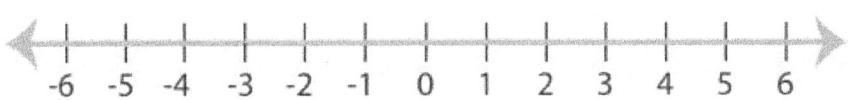

5) $x - 5 < -9$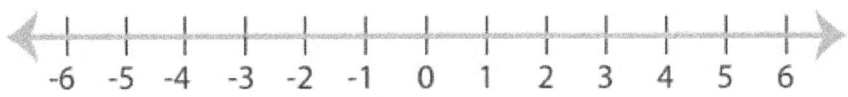

6) $9x \geq 72$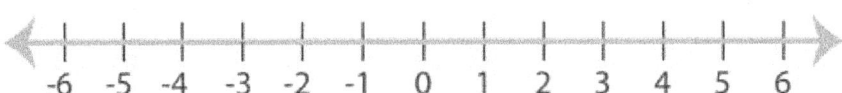

7) $9x \leq 27$

8) $x + 19 > 16$

WWW.MathNotion.Com

Multi-Step Inequalities

🍂 **Calculate each inequality.**

1) $x - 3 \leq 7$

2) $8 - x \leq 8$

3) $3x - 9 \leq 9$

4) $4x - 4 \geq 8$

5) $x - 7 \geq 1$

6) $5x - 15 \leq 5$

7) $6x - 8 \leq 4$

8) $-11 + 6x \leq 12$

9) $4(x - 4) \leq 16$

10) $3x - 10 \leq 11$

11) $5x - 25 < 25$

12) $9x - 5 < 22$

13) $20 - 7x \geq -15$

14) $33 + 6x < 45$

15) $8 + 8x \geq 96$

16) $7 + 3x < 13$

17) $4x - 3 < 9$

18) $5(2 - 2x) \geq -30$

19) $-(7 + 6x) < 29$

20) $12 - 8x \geq -20$

21) $-4(x - 6) > 24$

22) $\dfrac{3x + 9}{6} \leq 10$

23) $\dfrac{4x - 10}{3} \leq 2$

24) $\dfrac{2x - 8}{3} > 2$

25) $8 + \dfrac{x}{6} < 9$

26) $\dfrac{9x}{7} - 4 < 5$

27) $\dfrac{15x + 45}{15} > 1$

28) $16 + \dfrac{x}{4} < 6$

Systems of Equations

✏ **Calculate each system of equations.**

1) $-x + y = 2$ $x =$ ___
   $-4x + 2y = 6$ $y =$ ___

2) $-15x + 3y = -9$ $x =$ ___
   $9x - 16y = 48$ $y =$ ___

3) $y = -7$ $x =$ ___
   $6x + 5y = 7$ $y =$ ___

4) $3y = -9x + 15$ $x =$ ___
   $5x - 4y = -3$ $y =$ ___

5) $10x - 9y = -13$ $x =$ ___
   $-5x + 3y = 11$ $y =$ ___

6) $-12x - 16y = 20$ $x =$ ___
   $6x - 12y = 30$ $y =$ ___

7) $5x - 14y = -23$ $x =$ ___
   $-18x + 21y = 24$ $y =$ ___

8) $15x - 21y = -6$ $x =$ ___
   $2x - 3y = -2$ $y =$ ___

9) $-x + 3y = 3$ $x =$ ___
   $-14x + 16y = -10$ $y =$ ___

10) $x + 5y = 50$ $x =$ ___
    $3x + 10y = 80$ $y =$ ___

11) $6x - 7y = -8$ $x =$ ___
    $-x - 4y = -9$ $y =$ ___

12) $2x + 4y = -10$ $x =$ ___
    $2x - 8y = 14$ $y =$ ___

13) $4x + 3y = 12$ $x =$ ___
    $5x - 3y = 15$ $y =$ ___

14) $3x - 2y = 3$ $x =$ ___
    $7x - 8y = 22$ $y =$ ___

15) $3x + 2y = 5$ $x =$ ___
    $-10x - 4y = -14$ $y =$ ___

16) $10x + 7y = 1$ $x =$ ___
    $-5x - 7y = 24$ $y =$ ___

Common Core Subject Test Mathematics Grade 8

Systems of Equations Word Problems

✎ **Find the answer for each word problem.**

1) Tickets to a movie cost $4 for adults and $3 for students. A group of friends purchased 8 tickets for $31.00. How many adults ticket did they buy? _____

2) At a store, Eva bought two shirts and five hats for $77.00. Nicole bought three same shirts and four same hats for $84.00. What is the price of each shirt? _____

3) A farmhouse shelters 18 animals, some are pigs, and some are ducks. Altogether there are 66 legs. How many pigs are there? _____

4) A class of 214 students went on a field trip. They took 36 vehicles, some cars and some buses. If each car holds 5 students and each bus hold 22 students, how many buses did they take? _____

5) A theater is selling tickets for a performance. Mr. Smith purchased 5 senior tickets and 3 child tickets for $105 for his friends and family. Mr. Jackson purchased 3 senior tickets and 5 child tickets for $79. What is the price of a senior ticket? $_____

6) The difference of two numbers is 10. Their sum is 20. What is the bigger number? $_____

7) The sum of the digits of a certain two–digit number is 7. Reversing its digits increase the number by 9. What is the number? _____

8) The difference of two numbers is 11. Their sum is 25. What are the numbers? _____

9) The length of a rectangle is 5 meters greater than 2 times the width. The perimeter of rectangle is 28 meters. What is the length of the rectangle? _____

10) Jim has 25 nickels and dimes totaling $1.80. How many nickels does he have? _____

Common Core Subject Test Mathematics Grade 8

Answers of Worksheets

One–Step Equations

1) 30	9) 17	17) −14	25) 22
2) 7	10) −4	18) 20	26) 9
3) 4	11) 12	19) 45	27) 60
4) 6	12) 16	20) −13	28) 35
5) 5	13) 34	21) 34	29) 54
6) 11	14) −15	22) −5	30) 24
7) 11	15) −18	23) 47	31) 30
8) 8	16) 14	24) −42	32) −21

Multi–Step Equations

1) 2	9) 2	17) −2	25) 3
2) −7	10) −4	18) −7	26) 14
3) 4	11) −4	19) −5	27) 7
4) 8	12) −5	20) −11	28) 27
5) 6	13) −7	21) −4	29) −5
6) 2	14) 3	22) −2	30) 17
7) 10	15) −2	23) −7	31) −6
8) 2	16) 15	24) 6	32) −4

Graphing Single–Variable Inequalities

1)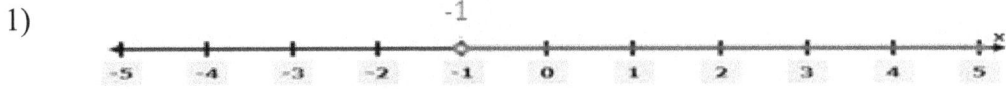

2)

3)

4)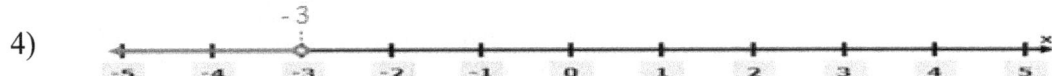

Common Core Subject Test Mathematics Grade 8

5) [number line with open circle at 1/2, shaded left]

6) [number line with closed circle at -2, shaded right]

7) [number line with closed circle at 3, shaded right]

8) [number line with open circle at -3.5, shaded right]

One–Step Inequalities

1) [number line with open circle at 0, shaded left]

2) [number line with closed circle at 7, shaded right]

3) [number line with open circle at 7, shaded right]

4) [number line with closed circle at 2, shaded left]

5) [number line with open circle at -4, shaded right]

6) [number line with closed circle at 8, shaded left]

7) [number line with closed circle at 3, shaded left]

8) [number line with open circle at -3, shaded right]

Multi-Step Inequalities

1) $x \leq 10$
2) $x \geq 0$
3) $x \leq 6$
4) $x \geq 3$
5) $x \geq 8$
6) $x \leq 4$
7) $x \leq 2$
8) $x \leq \frac{23}{6}$
9) $x \leq 8$
10) $x \leq 7$
11) $x < 10$
12) $x < 3$
13) $x \leq 5$
14) $x < 2$
15) $x \geq 11$
16) $x < 2$
17) $x < 3$
18) $x \leq 4$
19) $x > -6$
20) $x \leq 4$
21) $x < 0$
22) $x \leq 17$
23) $x \leq 4$
24) $x > 7$

WWW.MathNotion.Com

Common Core Subject Test Mathematics Grade 8

25) $x < 6$ 26) $x < 7$ 27) $x > -2$ 28) $x < -40$

Systems of Equations

1) $x = -1, y = 1$
2) $x = 0, y = -3$
3) $x = 7$
4) $x = 1, y = 2$
5) $x = -4, y = -3$
6) $x = 1, y = -2$
7) $x = 1, y = 2$
8) $x = 8, y = 6$
9) $x = 3, y = 2$
10) $x = -20, y = 14$
11) $x = 1, y = 2$
12) $x = -1, y = -2$
13) $x = 3, y = 0$
14) $x = -2, y = -\frac{9}{2}$
15) $x = 1, y = 1$
16) $x = 5, y = -7$

Systems of Equations Word Problems

1) 7
2) $16
3) 15
4) 2
5) $18
6) 15
7) 34
8) 18, 7
9) 11 meters
10) 14

Common Core Subject Test Mathematics Grade 8

Chapter 7 :
Linear Functions

Topics that you will practice in this chapter:

- ✓ Finding Slope
- ✓ Graphing Lines Using Line Equation
- ✓ Writing Linear Equations
- ✓ Graphing Linear Inequalities
- ✓ Finding Midpoint
- ✓ Finding Distance of Two Points

"Nature is written in mathematical language." – Galileo Galilei

Common Core Subject Test Mathematics Grade 8

Finding Slope

✎ **Find the slope of each line.**

1) $y = x + 8$

2) $y = -3x + 5$

3) $y = 2x + 12$

4) $y = -4x + 19$

5) $y = 11 + 6x$

6) $y = 7 - 5x$

7) $y = 8x + 19$

8) $y = -9x + 20$

9) $y = -7x + 4$

10) $y = 3x - 8$

11) $y = \frac{1}{3}x + 8$

12) $y = -\frac{4}{5}x + 9$

13) $-3x + 6y = 30$

14) $4x + 4y = 16$

15) $3y - x = 10$

16) $8y - x = 5$

✎ **Find the slope of the line through each pair of points.**

17) $(2, 3), (7, 10)$

18) $(-3, 5), (2, 15)$

19) $(5, -3), (1, 9)$

20) $(-5, -5), (10, 25)$

21) $(22, 3), (7, 18)$

22) $(-16, 8), (-7, 26)$

23) $(25, 11), (29, 19)$

24) $(26, -19), (14, 17)$

25) $(22, -13), (20, -11)$

26) $(19, 7), (15, -3)$

27) $(5, 7), (11, 19)$

28) $(52, -62), (40, 70)$

Common Core Subject Test Mathematics Grade 8

Graphing Lines Using Line Equation

✎ Sketch the graph of each line.

1) $y = x - 2$

2) $y = -3x + 2$

3) $x + y = 0$

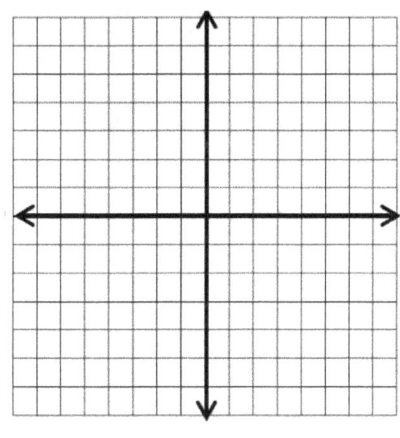
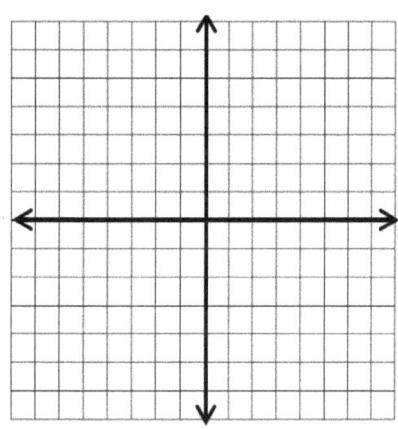
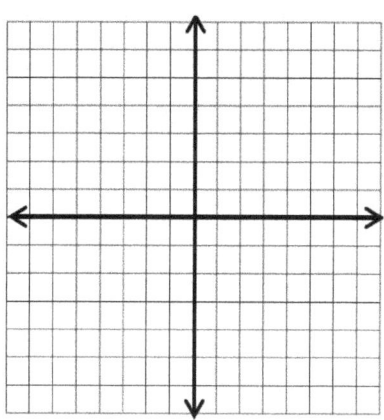

4) $x + y = -3$

5) $2x + 3y = -4$

6) $y - 3x + 6 = 0$

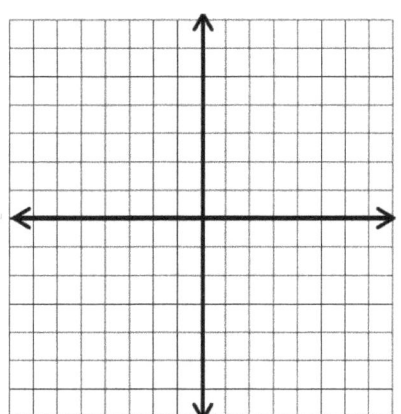
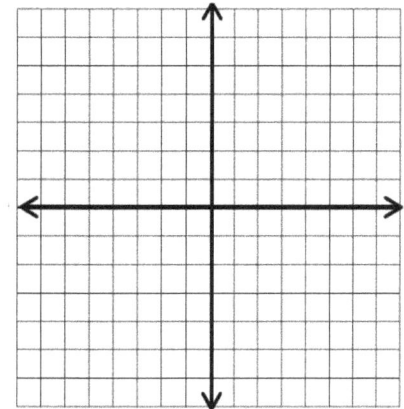
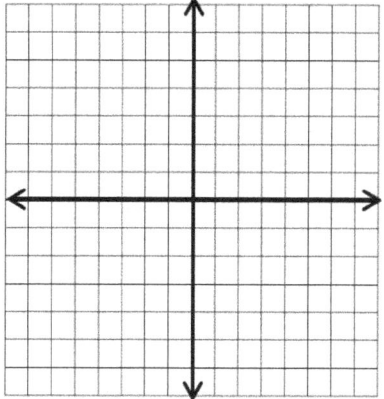

WWW.MathNotion.Com

Common Core Subject Test Mathematics Grade 8

Writing Linear Equations

✎ **Write the equation of the line through the given points.**

1) Through: $(2, -5), (3, 9)$

2) Through: $(-6, 3), (3, 12)$

3) Through: $(10, 7), (5, 27)$

4) Through: $(15, 11), (3, -1)$

5) Through: $(24, 17), (12, -7)$

6) Through: $(8, 29), (4, -7)$

7) Through: $(20, -16), (12, 0)$

8) Through: $(-3, 10), (2, -5)$

9) Through: $(-6, 17), (4, -3)$

10) Through: $(-8, 22), (5, -4)$

11) Through: $(9, 27), (3, -3)$

12) Through: $(11, 32), (9, 4)$

13) Through: $(-3, 13), (-4, 0)$

14) Through: $(-5, 5), (5, 15)$

15) Through: $(18, -32), (11, 3)$

16) Through: $(-4, 25), (4, -15)$

✎ **Find the answer for each problem.**

17) What is the equation of a line with slope 6 and intercept 12? _____

18) What is the equation of a line with slope -11 and intercept -4? _____

19) What is the equation of a line with slope -3 and passes through point $(5, 2)$? _____

20) What is the equation of a line with slope -5 and passes through point $(-2, -1)$? _____

21) The slope of a line is -10 and it passes through point $(-3, 0)$. What is the equation of the line? _____

22) The slope of a line is 8 and it passes through point $(0, 7)$. What is the equation of the line? _____

Graphing Linear Inequalities

✏️ **Sketch the graph of each linear inequality.**

1) $y > 4x - 5$
2) $y < 2x + 4$
3) $y \leq -5x - 2$

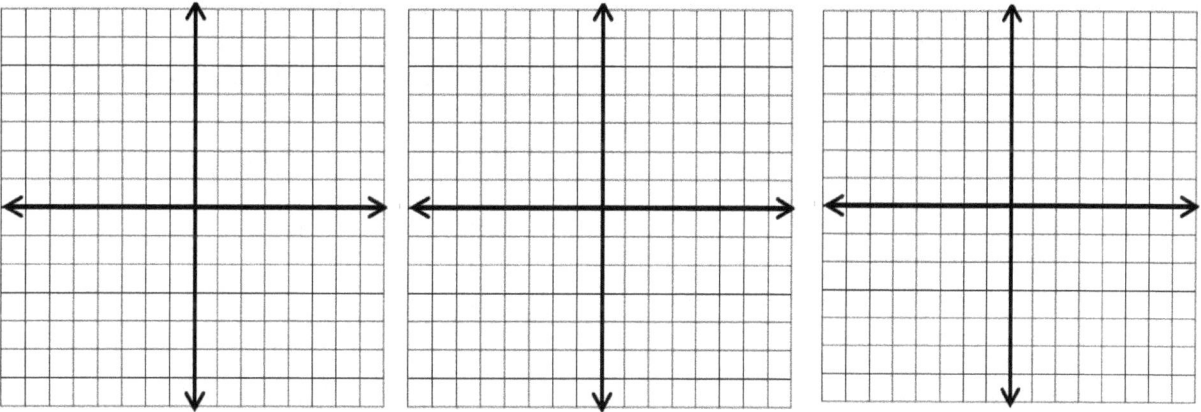

4) $4y \geq 12 + 4x$
5) $-12y < 3x - 24$
6) $5y \geq -15x + 10$

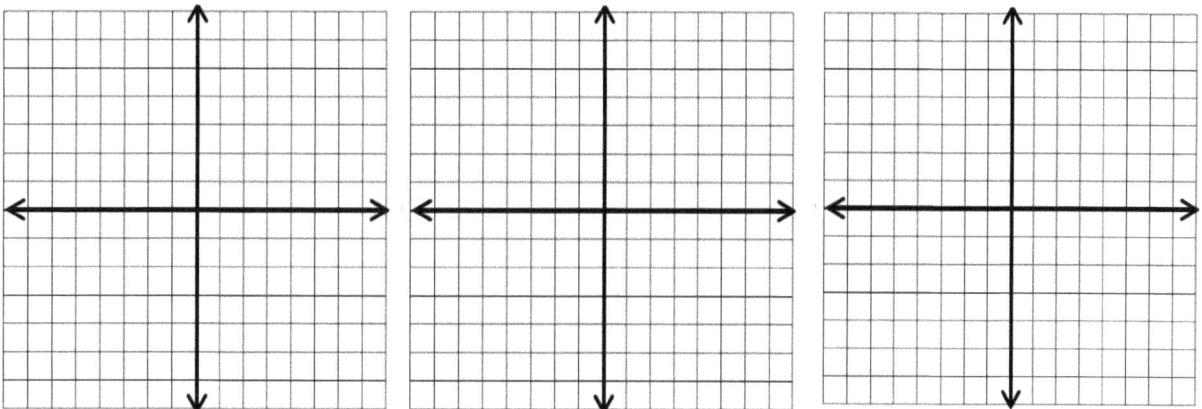

Common Core Subject Test Mathematics Grade 8

Finding Midpoint

✍ **Find the midpoint of the line segment with the given endpoints.**

1) $(-4, -3), (2, 3)$

2) $(9, 0), (-1, 8)$

3) $(9, -6), (3, 14)$

4) $(-10, -6), (0, 8)$

5) $(2, -5), (14, -15)$

6) $(-10, -3), (4, -13)$

7) $(8, 7), (-8, 13)$

8) $(-3, 6), (-9, 2)$

9) $(-4, 5), (16, -9)$

10) $(7, 14), (9, -2)$

11) $(-8, 6), (6, 6)$

12) $(10, 5), (-2, -3)$

13) $(-5, 12), (-3, 3)$

14) $(12, 7), (8, -2)$

15) $(10, 2), (-6, 14)$

16) $(-1, -2), (-7, 10)$

17) $(7, -7), (13, -13)$

18) $(-3, -8), (11, -4)$

19) $(5, -11), (-8, 9)$

20) $(14, -4), (16, 14)$

21) $(0, -5), (8, -1)$

22) $(3, 0), (-21, 18)$

23) $(17, -3), (-7, -5)$

24) $(26, -12), (6, 24)$

✍ **Find the answer for each problem.**

25) One endpoint of a line segment is $(-3, 7)$ and the midpoint of the line segment is $(-6, 9)$. What is the other endpoint? _____

26) One endpoint of a line segment is $(-3, 7)$ and the midpoint of the line segment is $(1, 5)$. What is the other endpoint? _____

27) One endpoint of a line segment is $(-10, -16)$ and the midpoint of the line segment is $(2, 9)$. What is the other endpoint? _____

Common Core Subject Test Mathematics Grade 8

Finding Distance of Two Points

✍ Find the distance between each pair of points.

1) $(6, 3), (-3, -9)$

2) $(5, 2), (-10, -6)$

3) $(8, 5), (8, 3)$

4) $(-8, -2), (2, 22)$

5) $(6, -7), (-3, -7)$

6) $(12, 0), (-9, -20)$

7) $(3, 20), (3, -5)$

8) $(10, 17), (5, 5)$

9) $(7, -2), (-4, -2)$

10) $(13, 4), (5, -2)$

11) $(11, 13), (5, 5)$

12) $(1, 4), (-23, -3)$

13) $(9, 8), (5, -4)$

14) $(-11, -4), (5, 8)$

15) $(-2, -6), (-2, -12)$

16) $(-1, -4), (23, 3)$

17) $(19, 3), (7, -6)$

18) $(-5, -2), (3, 4)$

19) $(2, 6), (2, -12)$

20) $(-4, -2), (8, -2)$

✍ Find the answer for each problem.

21) Triangle ABC is a right triangle on the coordinate system and its vertices are $(-2, 5)$, $(-2, 1)$, and $(1, 1)$. What is the area of triangle ABC? _____

22) Three vertices of a triangle on a coordinate system are $(3, -6)$, $(-5, -12)$, and $(3, -18)$. What is the perimeter of the triangle? _____

23) Four vertices of a rectangle on a coordinate system are $(-2, 2), (-2, 6)$, $(4, 2)$, and $(4, 6)$. What is its perimeter? _____

WWW.MathNotion.Com

Common Core Subject Test Mathematics Grade 8

Answers of Worksheets

Finding Slope

1) 1
2) -3
3) 2
4) -4
5) 6
6) -5
7) 8
8) -9
9) -7
10) 3
11) $\frac{1}{3}$
12) $-\frac{4}{5}$
13) $\frac{1}{2}$
14) -1
15) $\frac{1}{3}$
16) $\frac{1}{8}$
17) $\frac{7}{5}$
18) 2
19) -3
20) 2
21) -1
22) 2
23) 2
24) -3
25) -1
26) $\frac{5}{2}$
27) 2
28) -11

Graphing Lines Using Line Equation

1) $y = x - 2$

2) $y = -3x + 2$

3) $x + y = 0$

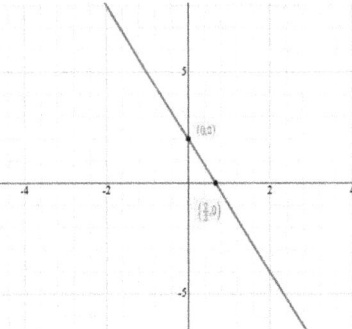
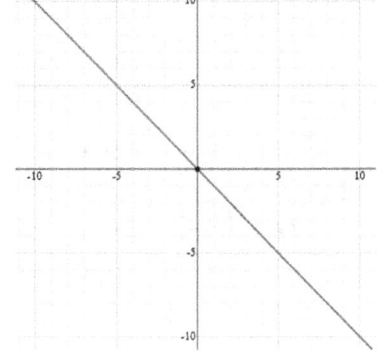

4) $x + y = -3$

5) $2x + 3y = -4$

6) $y - 3x + 6 = 0$

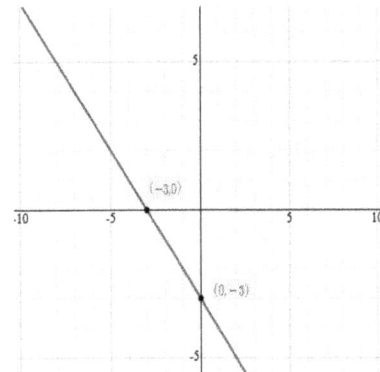
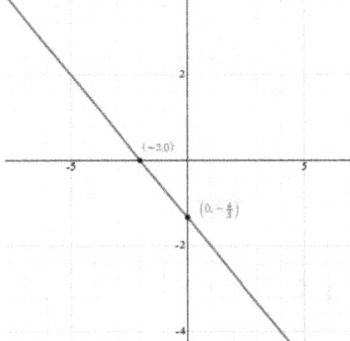
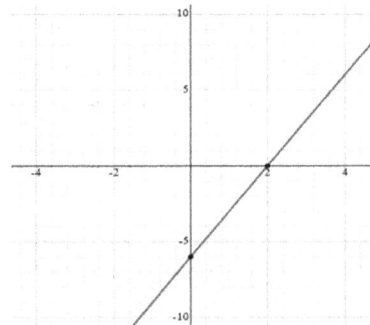

Common Core Subject Test Mathematics Grade 8

Writing Linear Equations

1) $y = 14x - 33$
2) $y = x + 9$
3) $y = -4x + 47$
4) $y = x - 4$
5) $y = 2x - 31$
6) $y = 9x - 43$
7) $y = -2x + 24$
8) $y = -3x + 1$
9) $y = -2x + 5$
10) $y = -2x + 6$
11) $y = 5x - 18$
12) $y = 14x - 122$
13) $y = 13x + 52$
14) $y = x + 10$
15) $y = -5x + 58$
16) $y = -5x + 5$
17) $y = 6x + 12$
18) $y = -11x - 4$
19) $y = -3x + 17$
20) $y = -5x - 11$
21) $y = -10x - 30$
22) $y = 8x + 7$

Graphing Linear Inequalities

1) $y > 4x - 5$

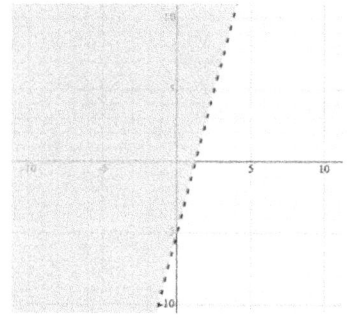

2) $y < 2x + 4$

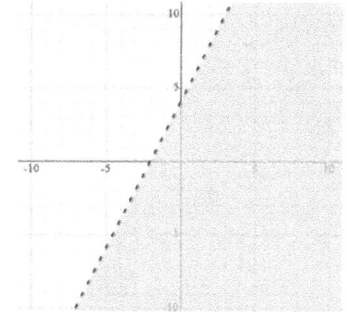

3) $y \leq -5x - 2$

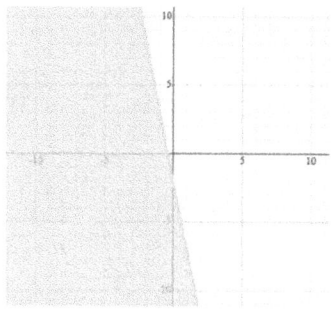

4) $4y \geq 12 + 4x$

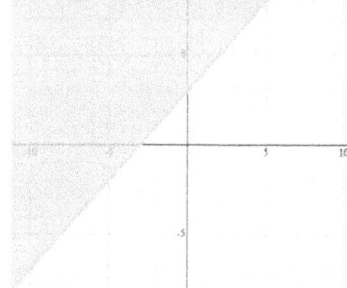

5) $-12y < 3x - 24$

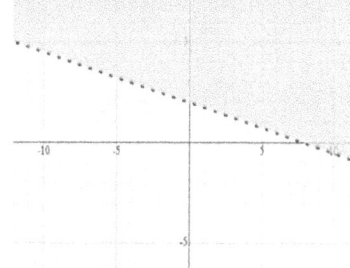

6) $5y \geq -15x + 10$

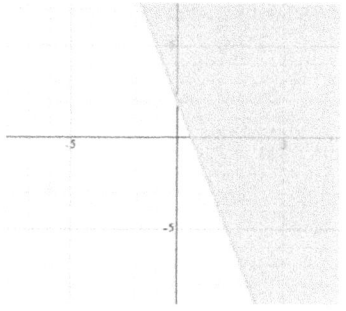

Finding Midpoint

1) $(-1, 0)$
2) $(4, 4)$
3) $(6, 4)$
4) $(-5, 1)$
5) $(8, -10)$
6) $(-3, -8)$
7) $(0, 10)$
8) $(-6, 4)$
9) $(6, -2)$
10) $(8, 6)$
11) $(-1, 6)$
12) $(4, 1)$
13) $(-4, 7.5)$
14) $(10, 2.5)$
15) $(2, 8)$
16) $(-4, 4)$
17) $(10, -10)$
18) $(4, -6)$

Common Core Subject Test Mathematics Grade 8

19) $(-1.5, -1)$
20) $(15, 5)$
21) $(4, -3)$

22) $(-9, 9)$
23) $(5, -4)$
24) $(16, 6)$

25) $(-9, 11)$
26) $(5, 3)$
27) $(14, 34)$

Finding Distance of Two Points

1) 15
2) 17
3) 2
4) 26
5) 9
6) 29
7) 25
8) 13

9) 11
10) 10
11) 10
12) 25
13) $4\sqrt{10}$
14) 20
15) 6
16) 25

17) 15
18) 10
19) 18
20) 12
21) 6 *square units*
22) 32 *units*
23) 20 *units*

Common Core Subject Test Mathematics Grade 8

Chapter 8:
Polynomials

Topics that you will practice in this chapter:

- ✓ Writing Polynomials in Standard Form
- ✓ Simplifying Polynomials
- ✓ Adding and Subtracting Polynomials
- ✓ Multiplying Monomials
- ✓ Multiplying and Dividing Monomials
- ✓ Multiplying a Polynomial and a Monomial
- ✓ Multiplying Binomials
- ✓ Factoring Trinomials
- ✓ Operations with Polynomials

Mathematics is the supreme judge; from its decisions there is no appeal. – Tobias Dantzig

Writing Polynomials in Standard Form

✎ Write each polynomial in standard form.

1) $11x - 7x =$

2) $-5 + 19x - 19x =$

3) $6x^5 - 12x^3 =$

4) $12 + 17x^4 - 12 =$

5) $5x^2 + 4x - 9x^3 =$

6) $-3x^2 + 12x^5 =$

7) $5x + 8x^3 - 2x^8 =$

8) $-7x^3 + 4x - 9x^6 =$

9) $3x^2 + 22 - 6x =$

10) $3 - 4x + 9x^4 =$

11) $13x^2 + 28x - 8x^3 =$

12) $16 + 4x^2 - 2x^3 =$

13) $19x^2 - 9x + 9x^4 =$

14) $3x^4 - 7x^2 - 2x^3 =$

15) $-51 + 3x^2 - 8x^4 =$

16) $7x^2 - 8x^6 + 4x^4 - 15 =$

17) $6x^4 - 4x^5 + 16 - 3x^3 =$

18) $-2x^6 + 4x - 7x^2 - 5x =$

19) $11x^7 + 8x^5 - 5x^7 - 3x^2 =$

20) $2x^2 - 12x^5 + 8x^2 + 3x^6 =$

21) $4x^5 - 11x^7 - 6x^3 + 16x^5 =$

22) $6x^3 + 3x^5 + 34x^4 - 8x^5 =$

23) $3x(4x + 5 - 2x^2) =$

24) $12x(x^6 + 4x^3) =$

25) $5x(3x^2 + 6x + 4) =$

26) $7x(4 - 2x + 6x^5) =$

27) $3x(4x^4 - 4x^3 + 2) =$

28) $4x(2x^5 + 6x^2 - 3) =$

29) $5x(3x^4 + 4x^3 + 2x) =$

30) $2x(3x - 2x^3 + 4x^6) =$

Common Core Subject Test Mathematics Grade 8

Simplifying Polynomials

✎ Simplify each expression.

1) $3(4x - 20) =$

2) $5x(3x - 4) =$

3) $6x(5x - 7) =$

4) $3x(7x + 5) =$

5) $5x(4x - 3) =$

6) $6x(8x + 2) =$

7) $(3x - 2)(x - 4) =$

8) $(x - 5)(2x + 6) =$

9) $(x - 3)(x - 7) =$

10) $(3x + 4)(3x - 4) =$

11) $(5x - 4)(5x - 2) =$

12) $6x^2 + 6x^2 - 8x^4 =$

13) $3x - 2x^2 + 5x^3 + 7 =$

14) $7x + 4x^2 - 10x^3 =$

15) $12x^2 + 5x^5 - 6x^3 =$

16) $-5x^2 + 4x^6 + 6x^8 =$

17) $-12x^3 + 10x^5 - 4x^6 + 4x =$

18) $11 - 7x^2 + 4x^2 - 16x^3 + 11 =$

19) $2x^2 - 9x + 4x^3 + 15x - 10x =$

20) $13 - 7x^5 + 6x^5 - 4x^2 + 5 =$

21) $-5x^8 + x^6 - 14x^3 + 5x^8 =$

22) $(7x^4 - 4) + (7x^4 - 2x^4) =$

23) $3(3x^4 - 4x^3 - 6x^4) =$

24) $-5(x^9 + 8) - 5(10 - x^9) =$

25) $8x^3 - 9x^4 - 2x + 19 - 8x^3 =$

26) $11 - 8x^3 + 6x^3 - 7x^5 + 6 =$

27) $(5x^3 - 4x) - (6x - 2 - 6x^3) =$

28) $4x^2 - 5x^4 - x(3x^3 + 2x) =$

29) $6x + 6x^5 - 10 - 4(x^5 - 3) =$

30) $4 - 3x^4 + (6x^5 - 2x^4 + 5x^5) =$

31) $-(x^5 + 4) - 8(3 + x^5) =$

32) $(4x^3 - 3x) - (3x - 5x^3) =$

WWW.MathNotion.Com

Common Core Subject Test Mathematics Grade 8

Adding and Subtracting Polynomials

✎ Add or subtract expressions.

1) $(-2x^2 - 3) + (3x^2 + 4) =$

2) $(4x^3 + 6) - (7 - 2x^3) =$

3) $(4x^5 + 5x^2) - (2x^5 + 15) =$

4) $(6x^3 - 2x^2) + (5x^2 - 4x) =$

5) $(10x^4 + 28x) - (34x^4 + 6) =$

6) $(7x^2 - 3) + (7x^2 + 3) =$

7) $(9x^2 + 4) - (10 - 5x^2) =$

8) $(6x^2 + x^5) - (x^5 + 4) =$

9) $(4x^3 - x) + (3x - 7x^3) =$

10) $(11x + 10) - (8x + 10) =$

11) $(15x^3 - 3x) - (3x - 4x^3) =$

12) $(4x - x^5) - (6x^5 + 8x) =$

13) $(2x^2 - 7x^7) - (4x^7 - 6x) =$

14) $(3x^2 - 5) + (8x^2 + 4x^5) =$

15) $(9x^4 + 5x^5) - (x^5 - 9x^4) =$

16) $(-4x^3 - 2x) + (9x - 5x^3) =$

17) $(4x - 3x^2) - (148x^2 + x) =$

18) $(5x - 8x^4) - (3x^4 - 4x^2) =$

19) $(8x^4 - 4) + (2x^4 - 3x^2) =$

20) $(5x^6 + 7x^3) - (x^3 - 5x^6) =$

21) $(-2x^2 + 20x^5 + 5x^4) + (12x^4 + 8x^5 + 24x^2) =$

22) $(7x^4 - 9x^7 - 6x) - (-3x^4 - 9x^7 + 6x) =$

23) $(14x + 12x^4 - 18x^6) + (20x^4 + 18x^6 - 10x) =$

24) $(5x^8 - 6x^6 - 4x) - (5x^3 + 9x^6 - 7x) =$

25) $(11x^2 - 6x^4 - 3x) - (-4x^2 - 12x^4 + 9x) =$

26) $(-5x^9 + 14x^3 + 3x^7) + (10x^7 + 26x^3 + 3x^9) =$

Common Core Subject Test Mathematics Grade 8

Multiplying Monomials

✎ Simplify each expression.

1) $6u^8 \times (-u^2) =$

2) $(-5p^8) \times (-2p^3) =$

3) $4xy^3z^5 \times 3z^4 =$

4) $3u^5t \times 8ut^4 =$

5) $(-5a^2) \times (-7a^3b^6) =$

6) $-3a^4b^3 \times 6a^2b =$

7) $13xy^5 \times x^4y^4 =$

8) $6p^4q^3 \times (-8pq^6) =$

9) $8s^4t^3 \times 4st^3 =$

10) $(-6x^4y^3) \times 6x^2y =$

11) $3xy^7z \times 12z^3 =$

12) $24xy \times x^2y =$

13) $13pq^4 \times (-3p^2q) =$

14) $13s^3t^4 \times st^4 =$

15) $11p^5 \times (-6p^3) =$

16) $(-8p^3q^5r) \times 3pq^4r^6 =$

17) $(-4a^4) \times (-7a^3b) =$

18) $6u^6v^2 \times (-5u^3v^4) =$

19) $9u^5 \times (-3u) =$

20) $-6xy^5 \times 4x^2y =$

21) $13y^5z^3 \times (-y^3z) =$

22) $8a^4bc^3 \times 2abc^3 =$

23) $(-7p^5q^6) \times (-5p^4q^2) =$

24) $4u^5v^3 \times (-4u^7v^3) =$

25) $17y^4z^5 \times (-y^6z) =$

26) $(-5pq^3r^2) \times 8p^2q^4r =$

27) $3ab^5c^6 \times 5a^4bc^2 =$

28) $6x^3yz^2 \times 3x^2y^7z^3 =$

WWW.MathNotion.Com

Common Core Subject Test Mathematics Grade 8

Multiplying and Dividing Monomials

✎ **Simplify each expression.**

1) $(5x^5)(2x^2) =$

2) $(4x^4)(6x^2) =$

3) $(3x^4)(7x^4) =$

4) $(5x^6)(4x^2) =$

5) $(12x^4)(3x^6) =$

6) $(4yx^8)(8y^4x^3) =$

7) $(14x^4y)(x^3y^5) =$

8) $(-5x^3y^4)(2x^3y^5) =$

9) $(-6x^4y^2)(-3x^3y^5) =$

10) $(5x^3y)(-5x^2y^3) =$

11) $(6x^4y^3)(4x^3y^4) =$

12) $(4x^3y^2)(5x^2y^4) =$

13) $(12x^3y^6)(4x^4y^{10}) =$

14) $(15x^3y^5)(3x^4y^6) =$

15) $(7x^2y^7)(8x^6y^7) =$

16) $(-3x^3y^8)(7x^9y^4) =$

17) $\dfrac{5x^6y^6}{xy^4} =$

18) $\dfrac{19x^7y^5}{19x^6y} =$

19) $\dfrac{56x^4y^4}{8xy} =$

20) $\dfrac{81x^5y^6}{9x^4y^5} =$

21) $\dfrac{36x^7y^6}{9x^2y^3} =$

22) $\dfrac{48x^9y^7}{4x^4y^6} =$

23) $\dfrac{88x^{18}y^{12}}{11x^8y^9} =$

24) $\dfrac{30x^7y^6}{6x^8y^3} =$

25) $\dfrac{150x^7y^6}{30x^4y^6} =$

26) $\dfrac{-42x^{18}y^{14}}{6x^4y^9} =$

27) $\dfrac{-36x^7y^8}{9x^5y^8} =$

WWW.MathNotion.Com

Multiplying a Polynomial and a Monomial

✏ Find each product.

1) $x(2x + 4) =$

2) $6(4 - 2x) =$

3) $5x(4x + 2) =$

4) $x(-4x + 5) =$

5) $8x(2x - 2) =$

6) $6(2x - 4y) =$

7) $7x(5x - 5) =$

8) $3x(12x + 2y) =$

9) $4x(x + 6y) =$

10) $11x(3x + 4y) =$

11) $7x(3x + 2) =$

12) $10x(4x - 10y) =$

13) $9x(3x - 2y) =$

14) $7x(x - 4y + 6) =$

15) $8x(2x^2 + 5y^2) =$

16) $12x(2x + 3y) =$

17) $4(2x^4 - 4y^4) =$

18) $4x(-3x^2y + 4y) =$

19) $-4(5x^3 - 2xy + 4) =$

20) $4(x^2 - 5xy - 6) =$

21) $8x(2x^3 - 5xy + 2x) =$

22) $-6x(-2x^3 - 6x + 2xy) =$

23) $3(2x^2 + xy - 9y^2) =$

24) $4x(5x^3 - 3x + 7) =$

25) $6(3x^{22} - 2x - 5) =$

26) $x^2(-2x^3 + 4x + 3) =$

27) $x^2(4x^3 + 10 - 2x) =$

28) $4x^4(3x^3 - 2x + 5) =$

29) $2x^2(4x^4 - 5xy + 7y^3) =$

30) $5x^2(5x^4 - 3x + 9) =$

31) $7x^2(6x^2 + 3x - 6) =$

32) $4x(x^3 - 4xy + 2y^2) =$

Multiplying Binomials

✏ **Find each product.**

1) $(x + 3)(x + 6) =$

2) $(x - 4)(x + 3) =$

3) $(x - 3)(x - 8) =$

4) $(x + 8)(x + 9) =$

5) $(x - 2)(x - 12) =$

6) $(x + 5)(x + 5) =$

7) $(x - 6)(x + 7) =$

8) $(x - 8)(x - 3) =$

9) $(x + 7)(x + 12) =$

10) $(x - 4)(x + 8) =$

11) $(x + 8)(x + 8) =$

12) $(x + 2)(x + 7) =$

13) $(x - 6)(x + 6) =$

14) $(x - 5)(x + 5) =$

15) $(x + 11)(x + 11) =$

16) $(x + 6)(x + 9) =$

17) $(x - 2)(x + 2) =$

18) $(x - 4)(x + 7) =$

19) $(3x + 5)(x + 6) =$

20) $(5x - 6)(4x + 8) =$

21) $(x - 7)(3x + 7) =$

22) $(x - 9)(x - 4) =$

23) $(x - 12)(x + 2) =$

24) $(2x - 4)(5x + 4) =$

25) $(3x - 8)(x + 8) =$

26) $(7x - 2)(6x + 3) =$

27) $(4x + 5)(3x + 5) =$

28) $(7x - 4)(9x + 4) =$

29) $(x + 2)(2x - 8) =$

30) $(5x - 4)(5x + 4) =$

31) $(3x + 2)(3x - 7) =$

32) $(x^2 + 8)(x^2 - 8) =$

Common Core Subject Test Mathematics Grade 8

Factoring Trinomials

✏️ **Factor each trinomial.**

1) $x^2 + 8x + 12 =$

2) $x^2 - 6x + 5 =$

3) $x^2 + 15x + 36 =$

4) $x^2 - 12x + 35 =$

5) $x^2 - 11x + 18 =$

6) $x^2 - 9x + 18 =$

7) $x^2 + 18x + 72 =$

8) $x^2 - x - 72 =$

9) $x^2 + 4x - 21 =$

10) $x^2 - 13x + 22 =$

11) $x^2 + 2x - 24 =$

12) $x^2 - 3x - 40 =$

13) $x^2 - 3x - 70 =$

14) $x^2 + 26x + 169 =$

15) $4x^2 - 7x - 15 =$

16) $x^2 - 14x + 33 =$

17) $10x^2 + 5x - 15 =$

18) $6x^2 - 4x - 42 =$

19) $x^2 + 12x + 36 =$

20) $5x^2 + 17x - 12 =$

✏️ **Calculate each problem.**

21) The area of a rectangle is $x^2 - x - 56$. If the width of rectangle is $x + 7$, what is its length? _____

22) The area of a parallelogram is $4x^2 + 17x - 15$ and its height is $x + 5$. What is the base of the parallelogram? _____

23) The area of a rectangle is $6x^2 - 22x + 12$. If the width of the rectangle is $3x - 2$, what is its length? _____

Common Core Subject Test Mathematics Grade 8

Operations with Polynomials

✎ **Find each product.**

1) $4(5x + 3) = $ _____

2) $8(2x + 6) = $ _____

3) $2(5x - 2) = $ _____

4) $-4(7x - 3) = $ _____

5) $3x^2(9x + 1) = $ _____

6) $4x^6(7x - 9) = $ _____

7) $3x^4(-7x + 3) = $ _____

8) $-8x^4(5x - 8) = $ _____

9) $7(x^2 + 5x - 3) = $ _____

10) $9(5x^2 - 7x + 5) = $ _____

11) $3(3x^2 + 3x + 2) = $ _____

12) $5x(3x^2 + 5x + 8) = $ _____

13) $(5x + 7)(3x - 3) = $ _____

14) $(9x + 3)(3x - 5) = $ _____

15) $(6x + 3)(4x - 2) = $ _____

16) $(7x - 2)(3x + 5) = $ _____

✎ **Calculate each problem.**

17) The measures of two sides of a triangle are $(2x + 5y)$ and $(6x - 3y)$. If the perimeter of the triangle is $(13x + 4y)$, what is the measure of the third side? _____

18) The height of a triangle is $(8x + 5)$ and its base is $(4x - 3)$. What is the area of the triangle? _____

19) One side of a square is $(6x + 2)$. What is the area of the square? _____

20) The length of a rectangle is $(5x - 8y)$ and its width is $(15x + 8y)$. What is the perimeter of the rectangle? _____

21) The side of a cube measures $(x + 2)$. What is the volume of the cube? _____

22) If the perimeter of a rectangle is $(28x + 6y)$ and its width is $(5x + 2y)$, what is the length of the rectangle? _____

WWW.MathNotion.Com

Common Core Subject Test Mathematics Grade 8

Answers of Worksheets

Writing Polynomials in Standard Form

1) $4x$
2) -5
3) $6x^5 - 12x^3$
4) $14x^4$
5) $-9x^3 + 5x^2 + 4x$
6) $12x^5 - 3x^2$
7) $-2x^8 + 8x^3 + 5x$
8) $-9x^6 - 7x^3 + 4x$
9) $3x^2 - 6x + 22$
10) $9x^4 - 4x + 3$
11) $-8x^3 + 13x^2 + 28x$
12) $-2x^3 + 4x^2 + 16$
13) $9x^4 + 19x^2 - 9x$
14) $3x^4 - 2x^3 - 7x^2$
15) $-8x^4 + 3x^2 - 51$
16) $-8x^6 + 4x^4 + 7x^2 - 15$
17) $-4x^5 + 6x^4 - 3x^3 + 16$
18) $-2x^6 - 7x^2 - x$
19) $6x^7 + 8x^5 - 3x^2$
20) $3x^6 - 12x^5 + 10x^2$
21) $-11x^7 + 20x^5 - 6x^3$
22) $-5x^5 + 34x^4 + 6x^3$
23) $-6x^3 + 12x^2 + 15x$
24) $12x^7 + 48x^4$
25) $15x^3 + 30x^2 + 20x$
26) $42x^6 - 14x^2 + 28x$
27) $12x^5 - 12x^4 + 6x$
28) $8x^6 + 24x^3 - 12x$
29) $15x^5 + 20x^4 + 10x^2$
30) $8x^7 - 4x^4 + 6x^2$

Simplifying Polynomials

1) $12x - 60$
2) $15x^2 - 20x$
3) $30x^2 - 42x$
4) $21x^2 + 15x$
5) $20x^2 - 15x$
6) $48x^2 + 12x$
7) $3x^2 - 14x + 8$
8) $2x^2 - 4x - 30$
9) $x^2 - 10x + 21$
10) $9x^2 - 16$
11) $25x^2 - 30x + 8$
12) $-8x^4 + 12x^2$
13) $5x^3 - 2x^2 + 3x + 7$
14) $-10x^3 + 4x^2 + 7x$
15) $5x^5 - 6x^3 + 12x^2$
16) $6x^8 + 4x^6 - 5x^2$
17) $-4x^6 + 10x^5 - 12x^3 + 4x$
18) $-16x^3 - 3x^2 + 22$
19) $4x^3 + 2x^2 - 4x$
20) $-x^5 - 4x^2 + 18$
21) $x^6 - 14x^3$
22) $12x^4 - 4$
23) $-9x^4 - 12x^3$
24) -90

Common Core Subject Test Mathematics Grade 8

25) $-9x^4 - 2x + 19$

26) $-7x^5 - 2x^3 + 17$

27) $11x^3 - 10x + 2$

28) $-8x^4 + 2x^2$

29) $2x^5 + 6x + 2$

30) $11x^5 - 5x^4 + 4$

31) $-9x^5 - 28$

32) $9x^3 - 6x$

Adding and Subtracting Polynomials

1) $x^2 + 1$

2) $6x^3 - 1$

3) $2x^5 + 5x^2 - 15$

4) $6x^3 + 3x^2 - 4x$

5) $-24x^4 + 28x - 6$

6) $14x^2$

7) $14x^2 - 6$

8) $6x^2 - 4$

9) $-3x^3 + 2x$

10) $3x$

11) $19x^3 - 6x$

12) $-7x^5 - 4x$

13) $-11x^7 + 2x^2 + 6x$

14) $4x^5 + 11x^2 - 5$

15) $4x^5 + 18x^4$

16) $-9x^3 + 7x$

17) $-151x^2 + 3x$

18) $-11x^4 + 4x^2 + 5x$

19) $10x^4 - 3x^2 - 4$

20) $10x^6 + 6x^3$

21) $28x^5 + 17x^4 + 22x^2$

22) $10x^4 - 12x$

23) $32x^4 + 4x$

24) $5x^8 - 15x^6 - 5x^3 + 3x$

25) $6x^4 + 15x^2 - 12x$

26) $-2x^9 + 13x^7 + 40x^3$

Multiplying Monomials

1) $-6u^{10}$

2) $10p^{11}$

3) $12xy^3z^9$

4) $24u^6t^5$

5) $35a^5b^6$

6) $-18a^6b^4$

7) $13x^5y^9$

8) $-48p^5q^9$

9) $32s^5t^6$

10) $-36x^6y^4$

11) $36xy^7z^4$

12) $24px^3y^2$

13) $-39p^3q^5$

14) $13s^4t^8$

15) $-66p^8$

16) $-24p^4q^9r^7$

17) $28a^7b$

18) $-30u^9v^6$

19) $-27u^6$

20) $-24x^3y^6$

21) $-13y^8z^4$

22) $16a^5b^2c^6$

23) $35p^9q^8$

24) $-16u^{12}v^6$

25) $-17y^{10}z^6$

26) $-40p^3q^7r^3$

27) $15a^5b^6c^8$

28) $18x^5y^8z^5$

Multiplying and Dividing Monomials

1) $10x^7$

2) $24x^6$

3) $21x^8$

4) $20x^8$

5) $36x^{10}$

6) $32x^{11}y^5$

7) $14x^7y^6$

8) $-10x^6y^9$

9) $18x^7y^7$

10) $-25x^5y^4$

11) $24x^7y^7$

12) $20x^5y^6$

WWW.MathNotion.Com

Common Core Subject Test Mathematics Grade 8

13) $48x^7y^{16}$
14) $45x^7y^{11}$
15) $56x^8y^{14}$
16) $-21x^{12}y^{12}$
17) $5x^5y^2$

18) xy^4
19) $7x^3y^3$
20) $9xy$
21) $4x^5y^3$
22) $12x^5y$

23) $8x^{10}y^3$
24) $5x^{-1}y^3$
25) $5x^3$
26) $-7x^{14}y^5$
27) $-4x^2$

Multiplying a Polynomial and a Monomial

1) $2x^2 + 4x$
2) $-12x + 24$
3) $20x^2 + 10x$
4) $-4x^2 + 5x$
5) $16x^2 - 16x$
6) $12x - 24y$
7) $35x^2 - 35x$
8) $36x^2 + 6xy$
9) $4x^2 + 24xy$
10) $33x^2 + 44xy$
11) $21x^2 + 14x$
12) $40x^2 - 100xy$
13) $27x^2 - 18xy$
14) $7x^2 - 28xy + 42x$
15) $16x^3 + 40xy^2$
16) $24x^2 + 36xy$

17) $8x^4 - 16y^4$
18) $-12x^3y + 16xy$
19) $-20x^3 + 8xy - 16$
20) $4x^2 - 20xy - 24$
21) $16x^4 - 40x^2y + 16x^2$
22) $12x^4 + 36x^2 - 12x^2y$
23) $6x^2 + 3xy - 27y^2$
24) $20x^4 - 12x^2 + 28x$
25) $18x^{22} - 12x - 30$
26) $-2x^5 + 4x^3 + 3x^2$
27) $4x^5 - 2x^3 + 10x^2$
28) $12x^7 - 8x^5 + 20x^4$
29) $8x^6 - 10x^3y + 14x^2y^3$
30) $25x^6 - 15x^3 + 45x^2$
31) $42x^4 + 21x^3 - 42x^2$
32) $4x^4 - 16x^2y + 8xy^2$

Multiplying Binomials

1) $x^2 + 9x + 18$
2) $x^2 - x - 12$
3) $x^2 - 11x + 24$
4) $x^2 + 17x + 72$
5) $x^2 - 14x + 24$
6) $x^2 + 10x + 25$
7) $x^2 + x - 42$

8) $x^2 - 11x + 24$
9) $x^2 + 19x + 84$
10) $x^2 + 4x - 32$
11) $x^2 + 16x + 64$
12) $x^2 + 9x + 14$
13) $x^2 - 36$
14) $x^2 - 25$

Common Core Subject Test Mathematics Grade 8

15) $x^2 + 22x + 121$
16) $x^2 + 15x + 54$
17) $x^2 - 4$
18) $x^2 + 3x - 28$
19) $3x^2 + 23x + 30$
20) $20x^2 + 16x - 48$
21) $3x^2 - 14x - 49$
22) $x^2 - 13x + 36$
23) $x^2 - 10x - 24$

24) $10x^2 - 12x - 16$
25) $3x^2 + 16x - 64$
26) $42x^2 + 9x - 6$
27) $12x^2 + 35x + 25$
28) $63x^2 - 8x - 16$
29) $2x^2 - 4x - 16$
30) $25x^2 - 16$
31) $9x^2 - 15x - 14$
32) $x^4 - 64$

Factoring Trinomials

1) $(x + 6)(x + 2)$
2) $(x - 5)(x - 1)$
3) $(x + 12)(x + 3)$
4) $(x - 5)(x - 7)$
5) $(x - 2)(x - 9)$
6) $(x - 6)(x - 3)$
7) $(x + 6)(x + 12)$
8) $(x + 8)(x - 9)$

9) $(x - 3)(x + 7)$
10) $(x - 11)(x - 2)$
11) $(x - 4)(x + 6)$
12) $(x - 8)(x + 5)$
13) $(x + 7)(x - 10)$
14) $(x + 13)(x + 13)$
15) $(4x + 5)(x - 3)$
16) $(x - 11)(x - 3)$

17) $(5x - 5)(2x + 3)$
18) $(2x - 6)(3x + 7)$
19) $(x + 6)(x + 6)$
20) $(5x - 3)(x + 4)$
21) $(x - 8)$
22) $(4x - 3)$
23) $(2x - 6)$

Operations with Polynomials

1) $20x + 12$
2) $16x + 48$
3) $10x - 4$
4) $-28x + 12$
5) $27x^3 + 3x^2$
6) $28x^7 - 36x^6$
7) $-21x^5 + 9x^4$
8) $-40x^5 + 64x^4$

9) $7x^2 + 35x - 21$
10) $45x^2 - 63x + 45$
11) $9x^2 + 9x + 6$
12) $15x^3 + 25x^2 + 40x$
13) $15x^2 + 6x - 21$
14) $27x^2 - 36x - 15$
15) $24x^2 - 6$
16) $21x^2 + 29x - 10$

17) $(5x + 2y)$
18) $16x^2 - 2x - \frac{15}{2}$
19) $36x^2 + 24x + 4$
20) $40x$
21) $x^3 + 6x^2 + 12x + 8$
22) $(9x + y)$

Common Core Subject Test Mathematics Grade 8

Chapter 9 :
Functions Operations and Quadratic

Topics that you will practice in this chapter:

- ✓ Evaluating Function
- ✓ Adding and Subtracting Functions
- ✓ Multiplying and Dividing Functions
- ✓ Composition of Functions
- ✓ Quadratic Equation
- ✓ Solving Quadratic Equations
- ✓ Quadratic Formula and the Discriminant
- ✓ Quadratic Inequalities
- ✓ Graphing Quadratic Functions
- ✓ Domain and Range of Radical Functions
- ✓ Solving Radical Equations

It's fine to work on any problem, so long as it generates interesting mathematics along the way – even if you don't solve it at the end of the day." – Andrew Wiles

Evaluating Function

✎ **Write each of following in function notation.**

1) $h = -8x + 3$

2) $k = 2a - 14$

3) $d = 11t$

4) $y = \frac{5}{12}x - \frac{7}{12}$

5) $m = 24n - 210$

6) $c = p^2 - 5p + 10$

✎ **Evaluate each function.**

7) $f(x) = 2x - 7$, find $f(-3)$

8) $g(x) = \frac{1}{9}x + 12$, find $f(18)$

9) $h(x) = -4x + 9$, find $f(3)$

10) $f(x) = -x + 19$, find $f(-3)$

11) $f(a) = 7a - 12$, find $f(3)$

12) $h(x) = 14 - 3x$, find $f(-4)$

13) $g(n) = 6n - 10$, find $f(2)$

14) $f(x) = -11x - 4$, find $f(-1)$

15) $k(n) = -20 - 3.5n$, find $f(2)$

16) $f(x) = -0.7x + 3.3$, find $f(-7)$

17) $g(n) = \frac{11n+8}{n}$, find $g(2)$

18) $g(n) = \sqrt{3n} + 12$, find $g(3)$

19) $h(x) = x^{-2} - 7$, find $h(\frac{1}{9})$

20) $h(n) = n^{-3} + 11$, find $h(\frac{1}{4})$

21) $h(n) = n^3 - 2$, find $h(\frac{1}{2})$

22) $h(n) = n^2 - 4$, find $h(-\frac{1}{3})$

23) $h(n) = 4n^2 - 13$, find $h(-5)$

24) $h(n) = -2n^3 - 6n$, find $h(2)$

25) $g(n) = \sqrt{16n^2} - \sqrt{n}$, find $g(4)$

26) $h(a) = \frac{-14a+9}{3a}$, find $h(-b)$

27) $k(a) = 12a - 14$, find $k(a - 3)$

28) $h(x) = \frac{1}{9}x + 18$, find $h(-18x)$

29) $h(x) = 8x^2 + 16$, find $h(\frac{x}{2})$

30) $h(x) = x^4 - 20$, find $h(-2x)$

Adding and Subtracting Functions

Perform the indicated operation.

1) $f(x) = 2x + 3$
   $g(x) = x + 7$
   Find $(f - g)(2)$

2) $g(a) = -5a - 8$
   $f(a) = -3a - 5$
   Find $(g - f)(-2)$

3) $h(t) = 4t + 3$
   $g(t) = 4t + 7$
   Find $(h - g)(t)$

4) $g(a) = -6a - 10$
   $f(a) = 3a^2 + 9$
   Find $(g - f)(x)$

5) $g(x) = \frac{5}{6}x - 23$
   $h(x) = \frac{5}{12}x + 25$
   Find $g(12) - h(12)$

6) $h(x) = \sqrt{3x} - 2$
   $g(x) = \sqrt{3x} + 5$
   Find $(h + g)(12)$

7) $f(x) = x^{-1}$
   $g(x) = x^2 + \frac{5}{x}$
   Find $(f - g)(-3)$

8) $h(n) = n^2 + 2$
   $g(n) = -4n + 6$
   Find $(h - g)(2a)$

9) $g(x) = -2x^2 - 5 - 4x$
   $f(x) = 7 + 2x$
   Find $(g - f)(3x)$

10) $g(t) = 11t - 4$
    $f(t) = -2t^2 + 5$
    Find $(g + f)(-t)$

11) $f(x) = 8x + 9$
    $g(x) = -5x^2 + 3x$
    Find $(f - g)(-x^2)$

12) $f(x) = -3x^4 - 5x$
    $g(x) = 2x^4 + 5x + 22$
    Find $(f + g)(3x^2)$

Multiplying and Dividing Functions

Perform the indicated operation.

1) $g(x) = -2x - 1$
   $f(x) = 4x + 3$
   Find $(g.f)(2)$

2) $f(x) = 5x$
   $h(x) = -2x + 3$
   Find $(f.h)(-2)$

3) $g(a) = 5a - 2$
   $h(a) = 2a - 3$
   Find $(g.h)(-3)$

4) $f(x) = 2x - 7$
   $h(x) = x - 5$
   Find $\left(\frac{f}{h}\right)(4)$

5) $f(x) = 8a^2$
   $g(x) = 3 + 2a$
   Find $\left(\frac{f}{g}\right)(2)$

6) $g(a) = \sqrt{4a} + 2$
   $f(a) = (-a)^4 + 1$
   Find $\left(\frac{g}{f}\right)(1)$

7) $g(t) = t^3 + 1$
   $h(t) = 5t - 2$
   Find $(g.h)(-2)$

8) $g(n) = n^2 + 2n - 4$
   $h(n) = -5n + 3$
   Find $(g.h)(1)$

9) $g(a) = (a - 3)^2$
   $f(a) = a^2 + 4$
   Find $\left(\frac{g}{f}\right)(3)$

10) $g(x) = -3x^2 + \frac{4}{5}x + 9$
    $f(x) = x^2 - 24$
    Find $\left(\frac{g}{f}\right)(5)$

11) $f(x) = 2x^3 - 5x^2 + 1$
    $g(x) = 3x - 1$
    Find $(f.g)(x)$

12) $f(x) = 5x - 2$
    $g(x) = x^3 - 2x$
    Find $(f.g)(x^2)$

Common Core Subject Test Mathematics Grade 8

Composition of Functions

Using $f(x) = 2x - 5$ and $g(x) = -2x$, find:

1) $f(g(2)) =$

2) $f(g(-1)) =$

3) $g(f(-4)) =$

4) $g(f(5)) =$

5) $f(g(3)) =$

6) $g(f(0)) =$

Using $f(x) = -\frac{1}{4}x + \frac{3}{4}$ and $g(x) = 2x^2$, find:

7) $g(f(-2)) =$

8) $g(f(4)) =$

9) $g(g(1)) =$

10) $f(f(1)) =$

11) $g(f(-4)) =$

12) $f(g(x)) =$

Using $f(x) = -2x + 2$ and $g(x) = x + 1$, find:

13) $g(f(1)) =$

14) $f(f(0)) =$

15) $f(g(-1)) =$

16) $f(g(-3)) =$

17) $g(f(2)) =$

18) $f(g(x)) =$

Using $f(x) = \sqrt{x + 9}$ and $g(x) = x - 9$, find:

19) $f(g(9)) =$

20) $g(f(-9)) =$

21) $f(g(4)) =$

22) $f(f(7)) =$

23) $g(f(-5)) =$

24) $g(g(0)) =$

Quadratic Equation

Multiply.

1) $(x-4)(x+6) = $ _____

2) $(x+5)(x+7) = $ _____

3) $(x-6)(x+8) = $ _____

4) $(x+2)(x-9) = $ _____

5) $(x-7)(x-8) = $ _____

6) $(3x+2)(x-3) = $ _____

7) $(4x-3)(x+2) = $ _____

8) $(4x-5)(x+1) = $ _____

9) $(7x+1)(x-6) = $ _____

10) $(5x+1)(3x-3) = $ _____

Factor each expression.

11) $x^2 - 2x - 8 = $ _____

12) $x^2 + 8x + 15 = $ _____

13) $x^2 - 2x - 24 = $ _____

14) $x^2 - 10x + 21 = $ _____

15) $x^2 + 10x + 21 = $ _____

16) $4x^2 + 9x + 5 = $ _____

17) $5x^2 + 13x - 6 = $ _____

18) $5x^2 + 17x - 12 = $ _____

19) $2x^2 + 7x + 5 = $ _____

20) $9x^2 - 21x + 6 = $ _____

Calculate each equation.

21) $(x+6)(x-3) = 0$

22) $(x+1)(x+8) = 0$

23) $(3x+6)(x+5) = 0$

24) $(2x-2)(4x+8) = 0$

25) $x^2 + x + 10 = 22$

26) $x^2 + 11x + 36 = 12$

27) $2x^2 + 9x + 9 = 5$

28) $x^2 + 3x - 24 = 4$

29) $5x^2 + 5x - 40 = 20$

30) $8x^2 + 8x = 48$

Solving Quadratic Equations

✎ Solve each equation by factoring or using the quadratic formula.

1) $(x+9)(x-1) = 0$

2) $(x+7)(x+6) = 0$

3) $(x-8)(x+3) = 0$

4) $(x-6)(x-4) = 0$

5) $(x+2)(x+12) = 0$

6) $(5x+4)(x+7) = 0$

7) $(6x+1)(4x+5) = 0$

8) $(2x+7)(x+8) = 0$

9) $(x+6)(3x+15) = 0$

10) $(12x+2)(x+8) = 0$

11) $x^2 = 8x$

12) $x^2 - 16 = 0$

13) $3x^2 + 6 = 9x$

14) $-2x^2 - 8 = 10x$

15) $5x^2 + 40x = 45$

16) $x^2 + 10x = 24$

17) $x^2 + 6x = 16$

18) $x^2 + 9x = -18$

19) $x^2 + 13x = -36$

20) $x^2 + 3x - 15 = 5x$

21) $x^2 + 8x + 7 = -8$

22) $3x^2 - 11x = -9 + x$

23) $10x^2 + 3 = 27x - 15$

24) $7x^2 - 6x + 8 = 8$

25) $2x^2 - 12 = -3x + 2$

26) $10x^2 - 26x - 3 = -15$

27) $3x^2 + 21 = -16x + 5$

28) $x^2 + 15x - 10 = -66$

29) $3x^2 - 8x - 8 = 4 + x$

30) $2x^2 + 6x - 24 = 12$

31) $3x^2 - 33x + 54 = -18$

32) $-10x^2 - 15x - 9 = -9 - 27x^2$

Common Core Subject Test Mathematics Grade 8

Quadratic Formula and the Discriminant

✎ Find the value of the discriminant of each quadratic equation.

1) $3x(x-8) = 0$

2) $2x^2 + 6x - 4 = 0$

3) $x^2 + 6x + 7 = 0$

4) $x^2 - x + 3 = 0$

5) $x^2 + 4x - 3 = 0$

6) $2x^2 + 6x - 10 = 0$

7) $3x^2 + 7x + 5 = 0$

8) $x^2 - 6x - 4 = 0$

9) $2x^2 + 8x + 3 = 0$

10) $x^2 + 7x - 5 = 0$

11) $5x^2 + 2x - 3 = 0$

12) $-3x^2 - 11x + 4 = 0$

13) $-6x^2 - 12x + 8 = 0$

14) $-x^2 - 9x - 12 = 0$

15) $7x^2 - 6x - 10 = 0$

16) $-4x^2 - 2x + 8 = 0$

17) $5x^2 + 8x - 2 = 0$

18) $6x^2 - 4x = 0$

19) $3x^2 - 5x + 2 = 0$

20) $4x^2 + 9x + 3 = 0$

✎ Find the discriminant of each quadratic equation then state the number of real and imaginary solutions.

21) $-4x^2 - 16 = 16x$

22) $20x^2 = 20x - 5$

23) $-11x^2 - 19x = 26$

24) $22x^2 - 4x + 1 = 18x^2$

25) $-11x^2 = -15x + 8$

26) $3x^2 + 6x + 9 = 6$

27) $13x^2 - 5x - 12 = -26$

28) $-8x^2 - 32x - 25 = 7$

WWW.MathNotion.Com

Graphing Quadratic Functions

✎ Sketch the graph of each function. Identify the vertex and axis of symmetry.

1) $y = (x + 3)^2 + 2$

2) $y = (x - 3)^2 - 2$

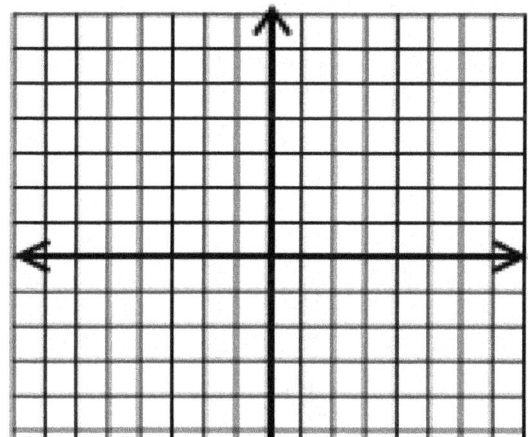
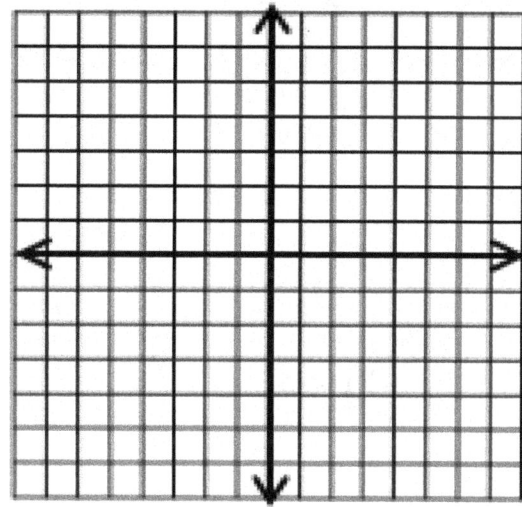

3) $y = 6 - (-x + 4)^2$

4) $y = -3x^2 - 6x + 9$

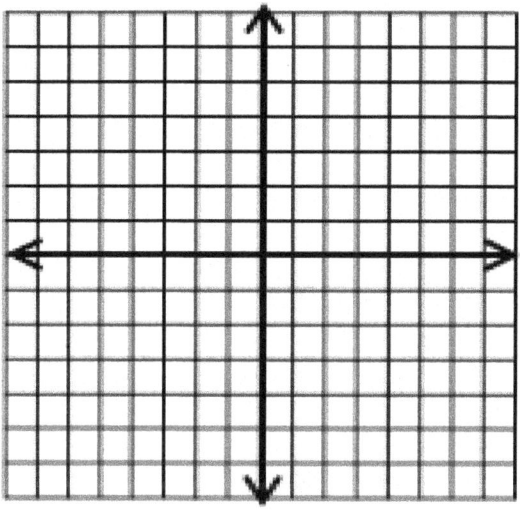
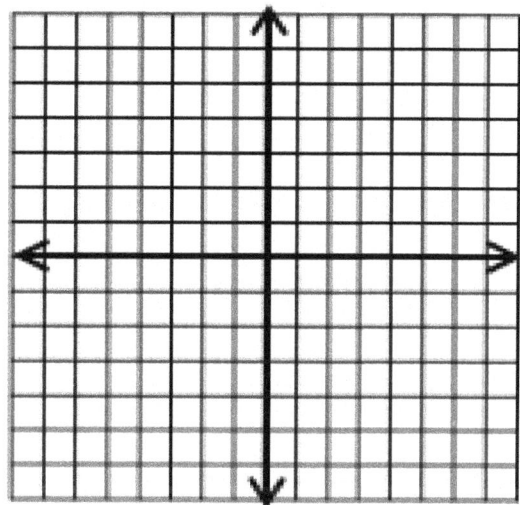

Quadratic Inequalities

✎ **Solve each quadratic inequality.**

1) $x^2 - 25 < 0$

2) $-x^2 - 6x - 8 > 0$

3) $5x^2 + 15x + 30 < 0$

4) $x^2 + 8x + 16 > 0$

5) $2x^2 - 18x - 20 \geq 0$

6) $x^2 > -10x - 25$

7) $3x^2 + 2x + 16 \leq 0$

8) $x^2 - 5x - 14 \leq 0$

9) $x^2 - 6x - 7 \geq 0$

10) $2x^2 + 16x - 18 < 0$

11) $x^2 + 6x - 72 > 0$

12) $3x^2 - 3x - 36 > 0$

13) $x^2 - 15x + 64 \leq 0$

14) $2x^2 - 24x + 72 \leq 0$

15) $x^2 - 16x + 63 \geq 0$

16) $x^2 - 16x + 55 \geq 0$

17) $x^2 - 81 \leq 0$

18) $x^2 - 17x + 42 \geq 0$

19) $9x^2 + 14x + 36 \leq 0$

20) $4x^2 - 2x - 24 > 2x^2$

21) $5x^2 - 20x + 20 < 0$

22) $7x^2 - 6x \geq 6x^2 - 5$

23) $5x^2 - 15 > 4x^2 + 2x$

24) $3x^2 - 4x \geq 3x^2 - 9x + 15$

25) $8x^2 + 9x - 54 > 5x^2$

26) $10x^2 + 50x - 60 < 0$

27) $-x^2 + 15x - 57 \geq 0$

28) $-5x^2 + 25x + 30 \leq 0$

29) $5x^2 + 40x + 75 < 0$

30) $9x^2 + 20x + 180 \leq 0$

31) $3x^2 + 2x - 36 \geq -x$

32) $3x^2 + 9x + 9 \leq 6x^2 + 3x$

Domain and Range of Radical Functions

✍ **Identify the domain and range of each function.**

1) $y = \sqrt{x+8} - 7$

2) $y = \sqrt[3]{3x-5} - 4$

3) $y = \sqrt{3x-9} + 3$

4) $y = \sqrt[3]{(4x+6)} - 2$

5) $y = 3\sqrt{4x+20} + 6$

6) $y = \sqrt[3]{(5x-2)} - 11$

7) $y = 4\sqrt{9x^2+8} + 3$

8) $y = \sqrt[3]{(7x^2-2)} - 6$

9) $y = 2\sqrt{2x^3+16} - 3$

10) $y = \sqrt[3]{(11x+4)} - 2x$

11) $y = 3\sqrt{-2(4x+8)} + 5$

12) $y = \sqrt[5]{(3x^2-12)} - 6$

13) $y = 3\sqrt{x-5} - 2$

14) $y = \sqrt[3]{6x+9} - 4$

✍ **Sketch the graph of each function.**

15) $y = -3\sqrt{x} + 5$

16) $y = 3\sqrt{x} - 6$

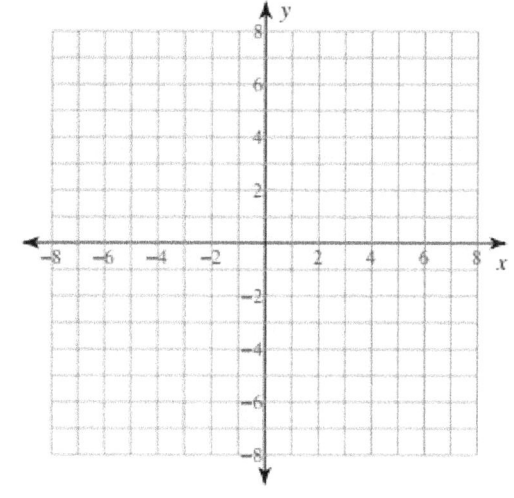
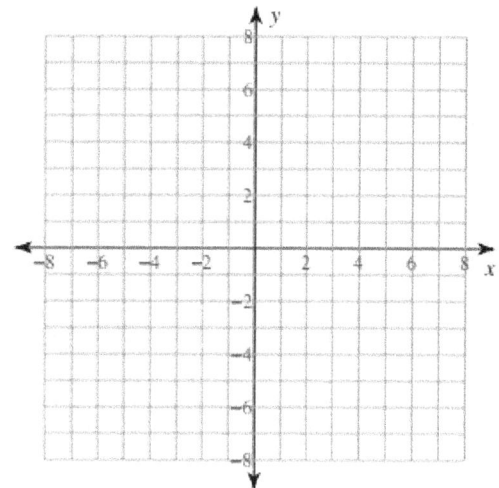

Common Core Subject Test Mathematics Grade 8

Solving Radical Equations

✎ Solve each equation. Remember to check for extraneous solutions.

1) $\sqrt{a} = 9$

2) $\sqrt{v} = 6$

3) $\sqrt{r} = 4$

4) $8 = 16\sqrt{x}$

5) $\sqrt{x+3} = 18$

6) $6 = \sqrt{x-7}$

7) $4 = \sqrt{r-3}$

8) $\sqrt{x-5} = 7$

9) $12 = \sqrt{x-4}$

10) $\sqrt{m+5} = 8$

11) $7\sqrt{5a} = 35$

12) $6\sqrt{2x} = 48$

13) $2 = \sqrt{6x-32}$

14) $\sqrt{304-4x} = 4$

15) $\sqrt{r+2} - 8 = 6$

16) $-21 = -7\sqrt{r+9}$

17) $60 = 6\sqrt{5v}$

18) $x = \sqrt{40-3x}$

19) $\sqrt{90-27a} = 3a$

20) $\sqrt{-8n+88} = 4$

21) $\sqrt{15r-5} = 4r-3$

22) $\sqrt{-64+32x} = 4x$

23) $\sqrt{4x+15} = \sqrt{2x+11}$

24) $\sqrt{12v} = \sqrt{15v-21}$

25) $\sqrt{9-x} = \sqrt{x-3}$

26) $\sqrt{6m+34} = \sqrt{8m+34}$

27) $\sqrt{7r+32} = \sqrt{-8-3r}$

28) $\sqrt{4k+10} = \sqrt{2-4k}$

29) $-20\sqrt{x-13} = -40$

30) $\sqrt{90-2x} = \sqrt{\dfrac{x}{4}}$

Common Core Subject Test Mathematics Grade 8

Answers of Worksheets

Evaluating Function

1) $h(x) = -8x + 3$
2) $k(a) = 2a - 14$
3) $d(t) = 11t$
4) $f(x) = \frac{5}{12}x - \frac{7}{12}$
5) $m(n) = 24n - 210$
6) $c(p) = p^2 - 5p + 10$
7) -13
8) 14
9) -3
10) 22
11) 9
12) 26
13) 2
14) 7
15) -27
16) 8.2
17) 15
18) 15
19) 74
20) 75
21) $-1\frac{7}{8}$
22) $-3\frac{8}{9}$
23) 87
24) -28
25) 14
26) $-\frac{14b+9}{3b}$
27) $12a - 50$
28) $-2x + 18$
29) $2x^2 + 16$
30) $16x^4 - 20$

Adding and Subtracting Functions

1) -2
2) 1
3) -4
4) $-3x^2 - 6x - 19$
5) -43
6) 15
7) $-7\frac{2}{3}$
8) $4a^2 + 8a - 4$
9) $-18x^2 - 18x - 12$
10) $-2t^2 - 11t + 1$
11) $5x^4 - 5x^2 + 9$
12) $-81x^8 + 22$

Multiplying and Dividing Functions

1) -55
2) -70
3) 153
4) -1
5) $4\frac{4}{7}$
6) 2
7) 84
8) 2
9) 0
10) -62
11) $6x^4 - 17x^3 + 5x^2 + 3x - 1$
12) $5x^8 - 2x^6 - 10x^4 + 4x^2$

Composition of Functions

1) -13
2) -1
3) 26
4) -10
5) -17
6) 10
7) $\frac{25}{8}$
8) $\frac{1}{8}$
9) 8
10) $\frac{5}{8}$
11) $\frac{49}{8}$
12) $-\frac{1}{2}(x^2 - \frac{3}{2})$
13) 1
14) -2
15) 2
16) 6
17) -1
18) $-2x$

WWW.MathNotion.Com

Common Core Subject Test Mathematics Grade 8

19) 3
20) −9

21) 2
22) $\sqrt{13}$

23) −7
24) −18

Quadratic Equations

1) $x^2 + 2x - 24$
2) $x^2 + 12x + 35$
3) $x^2 + 2x - 48$
4) $x^2 - 7x - 18$
5) $x^2 - 15x + 56$
6) $3x^2 - 7x - 6$
7) $4x^2 + 5x - 6$
8) $4x^2 - x - 5$
9) $7x^2 - 41x - 6$
10) $15x^2 - 12x - 3$

11) $(x-4)(x+2)$
12) $(x+5)(x+3)$
13) $(x-6)(x+4)$
14) $(x-3)(x-7)$
15) $(x+3)(x+7)$
16) $(4x+5)(x+1)$
17) $(5x-2)(x+3)$
18) $(5x-3)(x+4)$
19) $(2x+5)(x+1)$
20) $3(x-2)(3x-1)$

21) $x = -6, x = 3$
22) $x = -1, x = -8$
23) $x = -2, x = -5$
24) $x = 1, x = -2$
25) $x = 3, x = -4$
26) $x = -3, x = -8$
27) $x = -4, x = -\frac{1}{2}$
28) $x = 4, x = -7$
29) $x = 3, x = -4$
30) $x = -3, x = 2$

Solving quadratic equations

1) $\{-9, 1\}$
2) $\{-6, -7\}$
3) $\{8, -3\}$
4) $\{6, 4\}$
5) $\{-2, -12\}$
6) $\{-\frac{4}{5}, -7\}$
7) $\{-\frac{5}{4}, -\frac{1}{6}\}$
8) $\{-\frac{7}{2}, -8\}$

9) $\{-6, -5\}$
10) $\{-\frac{1}{6}, -8\}$
11) $\{8, 0\}$
12) $\{4, -4\}$
13) $\{2, 1\}$
14) $\{-4, -1\}$
15) $\{1, -9\}$
16) $\{2, -12\}$

17) $\{2, -8\}$
18) $\{-3, -6\}$
19) $\{-4, -9\}$
20) $\{5, -3\}$
21) $\{-5, -3\}$
22) $\{1, 3\}$
23) $\{\frac{6}{5}, \frac{3}{2}\}$
24) $\{\frac{6}{7}, 0\}$

25) $\{-\frac{7}{2}, 2\}$
26) $\{\frac{3}{5}, 2\}$
27) $\{-\frac{4}{3}, -4\}$
28) $\{-8, -7\}$
29) $\{4, -1\}$
30) $\{3, -6\}$
31) $\{3, 8\}$
32) $\{\frac{15}{17}, 0\}$

Quadratic formula and the discriminant

1) 576
2) 68
3) 8
4) −11
5) 28

6) 116
7) −11
8) 52
9) 40
10) 69

11) 64
12) 169
13) 336
14) 33
15) 316

16) 132
17) 104
18) 16
19) 1
20) 33

21) 0, one real solution
22) 0, one real solution
23) −783, no solution

WWW.MathNotion.Com

Common Core Subject Test Mathematics Grade 8

24) 0, one real solution 26) 0, one real solution 28) 0, one real solution
25) −127, no solution   27) −703, no solution

Graphing quadratic functions

1) $(-3, 2), x = -3$

2) $(3, -2), x = 3$

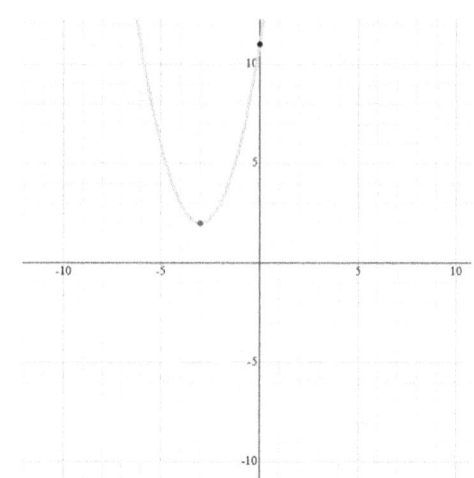

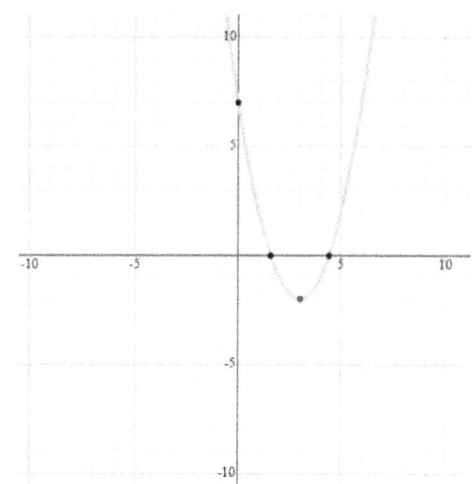

3) $(4, 6), x = 4$

4) $(-1, 12), x = -1$

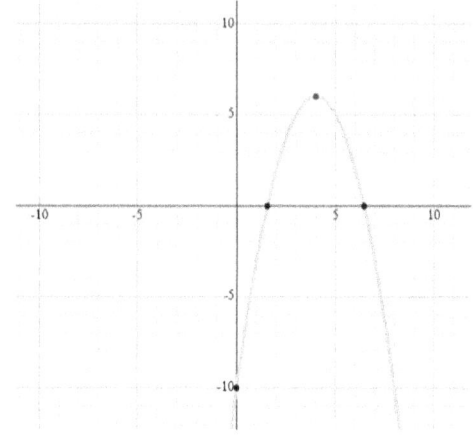

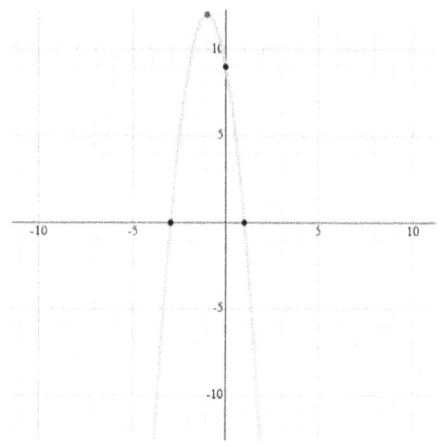

Quadratic inequalities

1) $-5 < x < 5$
2) $-4 < x < -2$
3) no solution
4) $x < -4 \text{ or } x > -4$
5) $x \leq -1 \text{ or } x \geq 10$
6) $x < -5 \text{ or } x > -5$
7) no solution
8) $-2 \leq x \leq 7$
9) $x \leq -1 \text{ or } x \geq 7$
10) $-9 < x < 1$
11) $x < -12 \text{ or } x > 6$
12) $-3 < x < 4$
13) no solution
14) $x = 6$
15) $x \leq 7 \text{ or } x \geq 9$
16) $x \leq 5 \text{ or } x \geq 11$
17) $-9 \leq x \leq 9$
18) $x \leq 3 \text{ or } x \geq 14$

Common Core Subject Test Mathematics Grade 8

19) no solution
20) $x < -3$ or $x > 4$
21) no solution
22) $x \leq 1$ or $x \geq 5$
23) $x < -3$ or $x > 5$

24) $x \geq 3$
25) $x < -6$ or $x > 3$
26) $-6 < x < 1$
27) no solution
28) $x \leq -1$ or $x \geq 6$

29) $-5 < x < -3$
30) no solution
31) $x \leq -4$ or $x \geq 3$
32) $x \leq -1$ or $x \geq 3$

Domain and range of radical functions

1) domain: $x \geq -8$
   range: $y \geq -7$
2) domain: {all real numbers}
   range: {all real numbers}
3) domain: $x \geq 3$
   range: $y \geq 3$
4) domain: {all real numbers}
   range: {all real numbers}
5) domain: $x \geq -5$
   range: $y \geq 6$
6) domain: {all real numbers}
   range: {all real numbers}
7) domain: {all real numbers}
   range: $y \geq 8\sqrt{2} + 3$

8) domain: {all real numbers}
   range: {all real numbers}
9) domain: $x \geq -2$
   range: $y \geq -3$
10) domain: {all real numbers}
    range: {all real numbers}
11) domain: $x \leq -2$
    range: $y \geq 5$
12) domain: {all real numbers}
    range: {all real numbers}
13) domain: $x \geq 5$
    range: $y \geq -2$
14) domain: {all real numbers}
    range: {all real numbers}

15)

16)

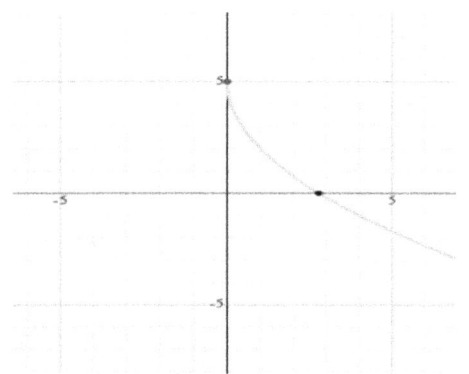

Common Core Subject Test Mathematics Grade 8

Solving radical equations

1) {81}
2) {36}
3) {16}
4) {$\frac{1}{4}$}
5) {321}
6) {43}
7) {19}
8) {54}
9) {148}
10) {59}

11) {5}
12) {32}
13) {6}
14) {72}
15) {194}
16) {0}
17) {20}
18) {5}
19) {2}
20) {9}

21) {2}
22) no solution
23) {−2}
24) {7}
25) {6}
26) {0}
27) {−4}
28) {−1}
29) {17}
30) {40}

Common Core Subject Test Mathematics Grade 8

Chapter 10:
Geometry and Solid Figures

Topics that you will practice in this chapter:

- ✓ Angles
- ✓ Pythagorean Relationship
- ✓ Triangles
- ✓ Polygons
- ✓ Trapezoids
- ✓ Circles
- ✓ Cubes
- ✓ Rectangular Prism
- ✓ Cylinder
- ✓ Pyramids and Cone

Mathematics is, as it were, a sensuous logic, and relates to philosophy as do the arts, music, and plastic art to poetry. — *K. Shegel*

Common Core Subject Test Mathematics Grade 8

Angles

✎ What is the value of x in the following figures?

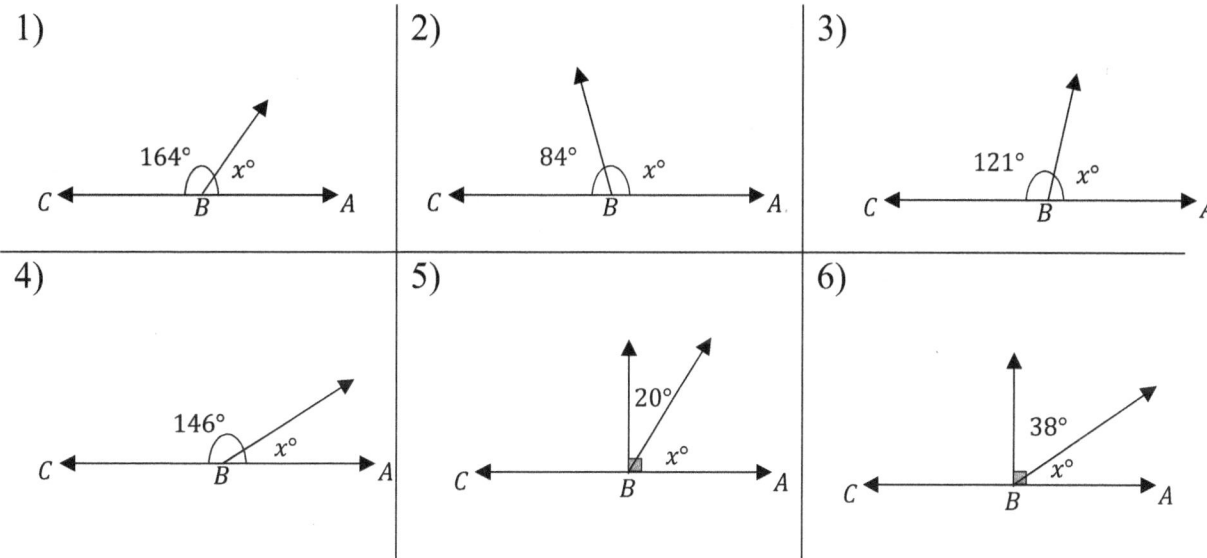

✎ Calculate.

7) Two supplement angles have equal measures. What is the measure of each angle? _____

8) The measure of an angle is seven fifth the measure of its supplement. What is the measure of the angle? _____

9) Two angles are complementary and the measure of one angle is 24 less than the other. What is the measure of the smaller angle? _____

10) Two angles are complementary. The measure of one angle is one fifth the measure of the other. What is the measure of the bigger angle? _____

11) Two supplementary angles are given. The measure of one angle is 40° less than the measure of the other. What does the smaller angle measure? _____

Common Core Subject Test Mathematics Grade 8

Pythagorean Relationship

✎ **Do the following lengths form a right triangle?**

1)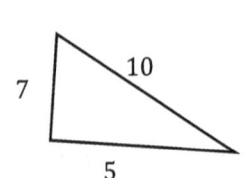

2)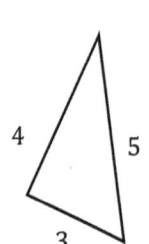

3)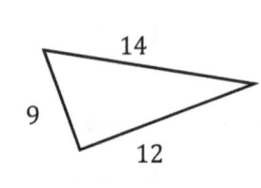

4)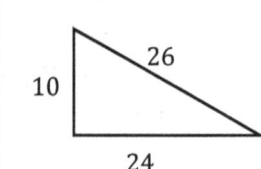

5)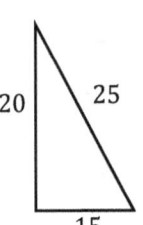

6)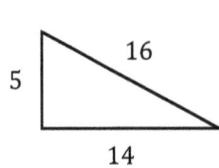

7)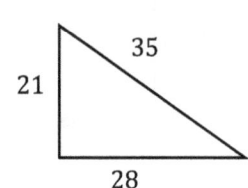

8)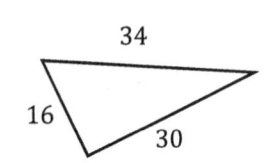

✎ **Find the missing side?**

9)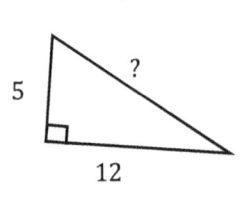

10)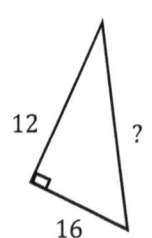

11)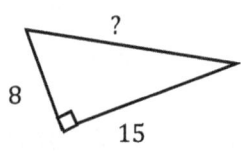

12)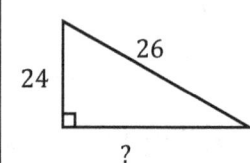

13)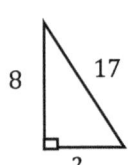

14)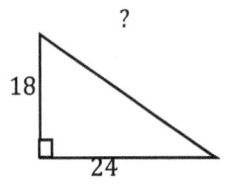

15)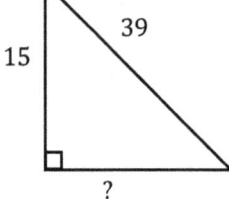

16)

WWW.MathNotion.Com

Triangles

✎ Find the measure of the unknown angle in each triangle.

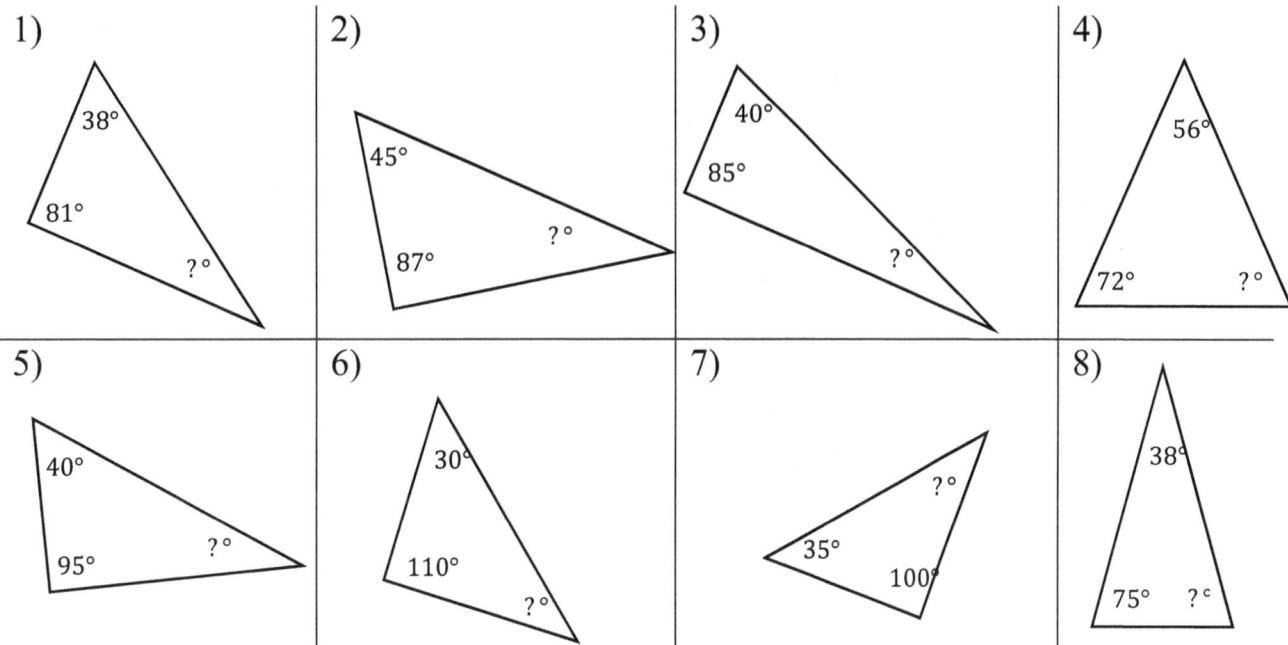

✎ Find area of each triangle.

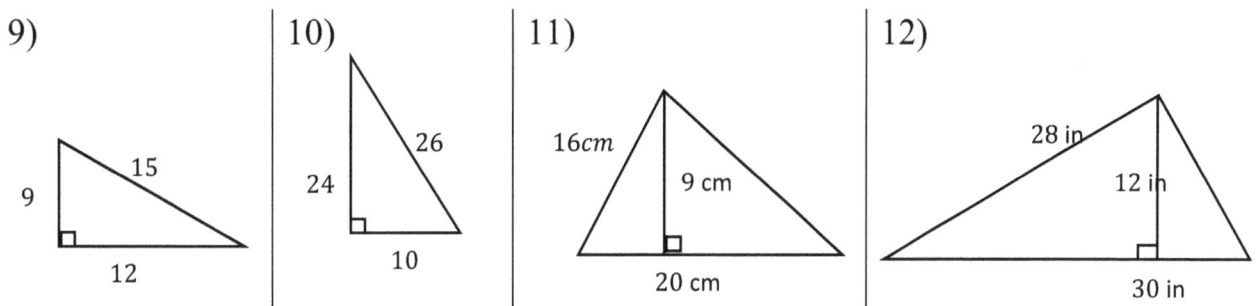

Common Core Subject Test Mathematics Grade 8

Polygons

✏️ **Find the perimeter of each shape.**

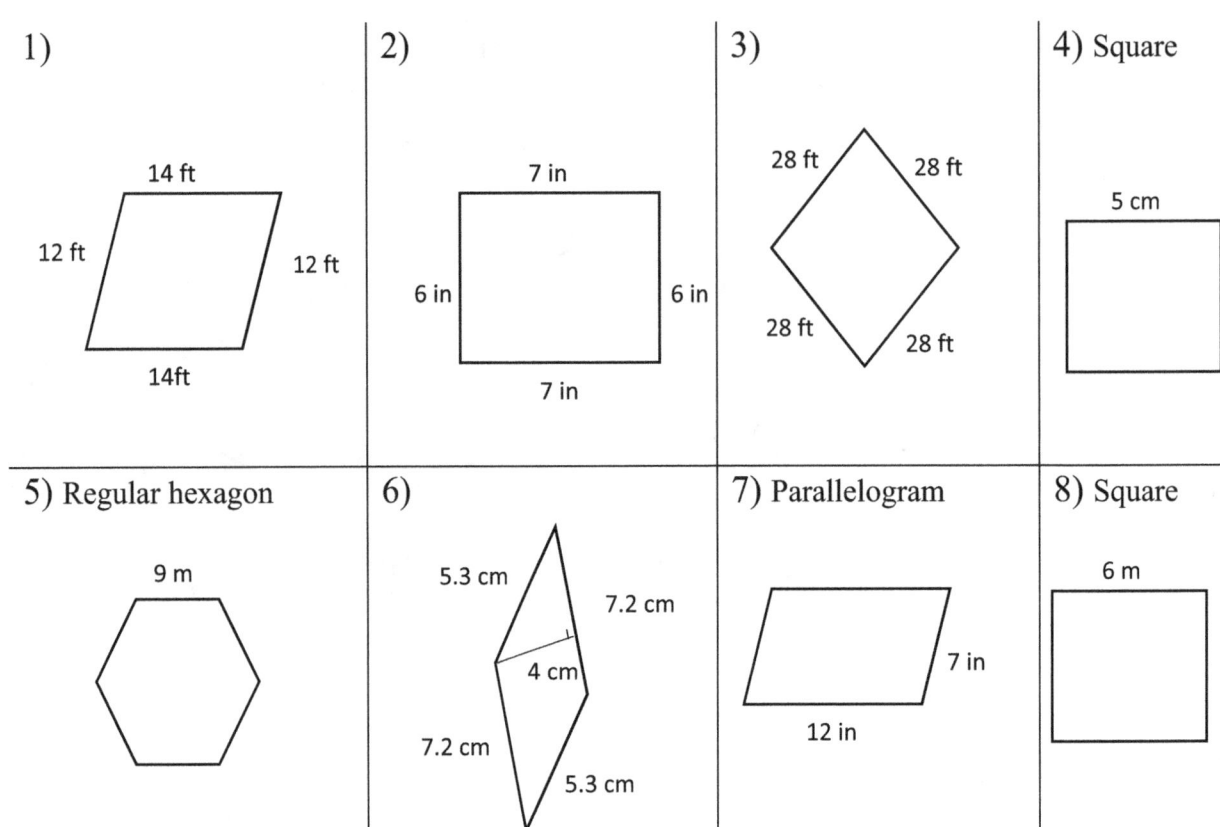

✏️ **Find the area of each shape.**

9) Parallelogram — 5 m, 6 m, 5 m

10) Rectangle — 25 in, 12 in

11) Rectangle — 16 km, 10 km

12) Square — 7 in

Common Core Subject Test Mathematics Grade 8

Trapezoids

✎ Find the area of each trapezoid.

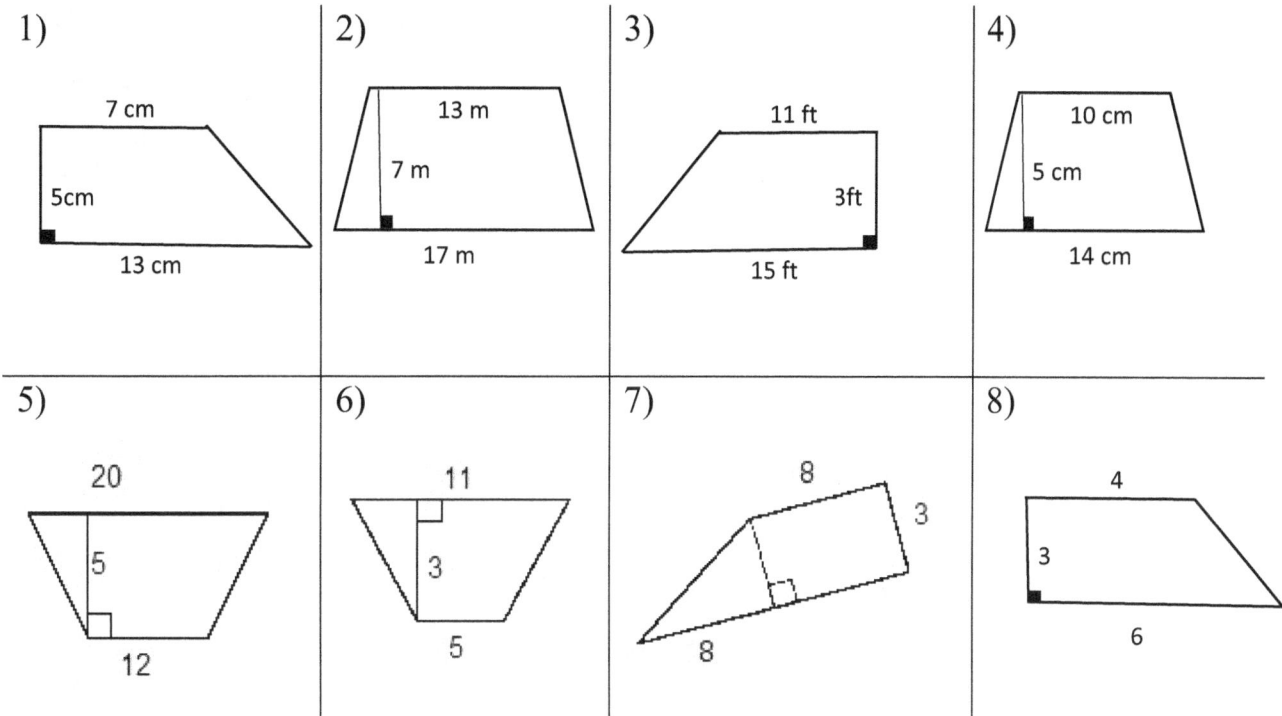

✎ Calculate.

1) A trapezoid has an area of 45 cm² and its height is 5 cm and one base is 5 cm. What is the other base length? _____

2) If a trapezoid has an area of 99 ft² and the lengths of the bases are 8 ft and 10 ft, find the height? _____

3) If a trapezoid has an area of 126 m² and its height is 14 m and one base is 6 m, find the other base length? _____

4) The area of a trapezoid is 440 ft² and its height is 22 ft. If one base of the trapezoid is 15 ft, what is the other base length?

_____

Circles

✎ **Find the area of each circle.** ($\pi = 3.14$)

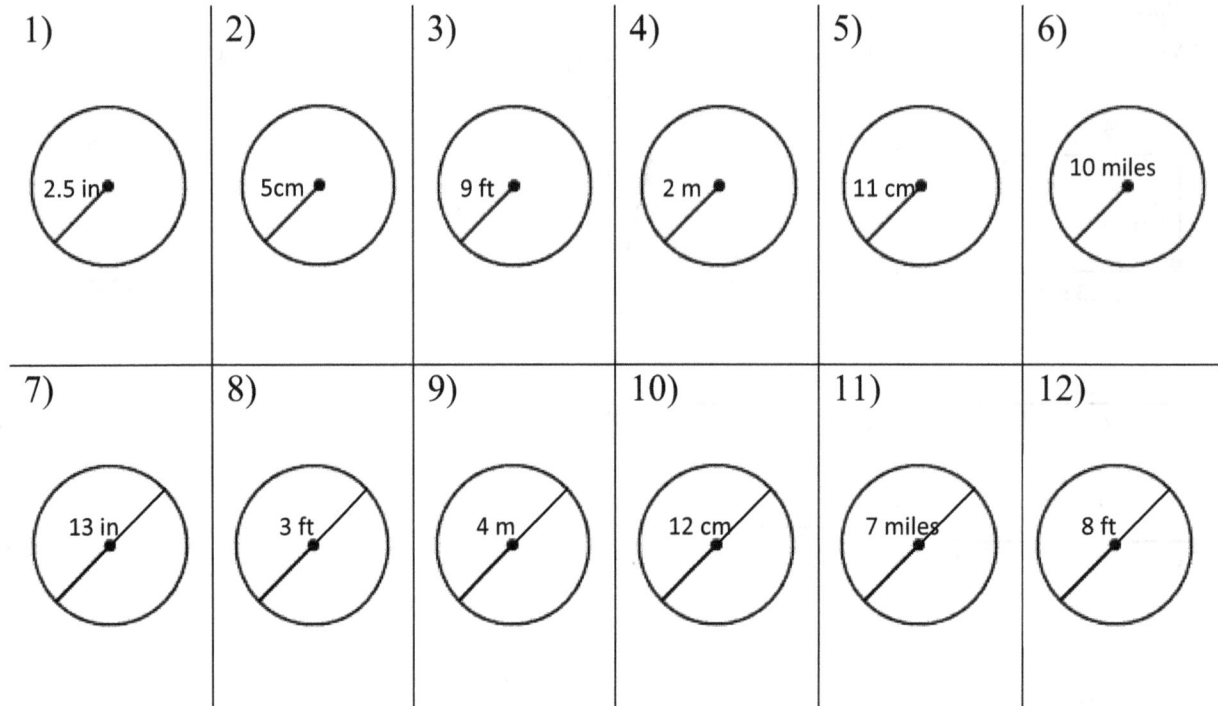

1) 2.5 in
2) 5 cm
3) 9 ft
4) 2 m
5) 11 cm
6) 10 miles
7) 13 in
8) 3 ft
9) 4 m
10) 12 cm
11) 7 miles
12) 8 ft

✎ **Complete the table below.** ($\pi = 3.14$)

Circle No.	Radius	Diameter	Circumference	Area
1	1 in	2 in	6.28 in	3.14 in^2
2		10 m		
3				28.26 ft^2
4			47.1 mi	
5		11 km		
6	7 cm			
7		12 ft		
8				314 m^2
9			56.52 in	
10	4.5 ft			

Common Core Subject Test Mathematics Grade 8

Cubes

✏ Find the volume of each cube.

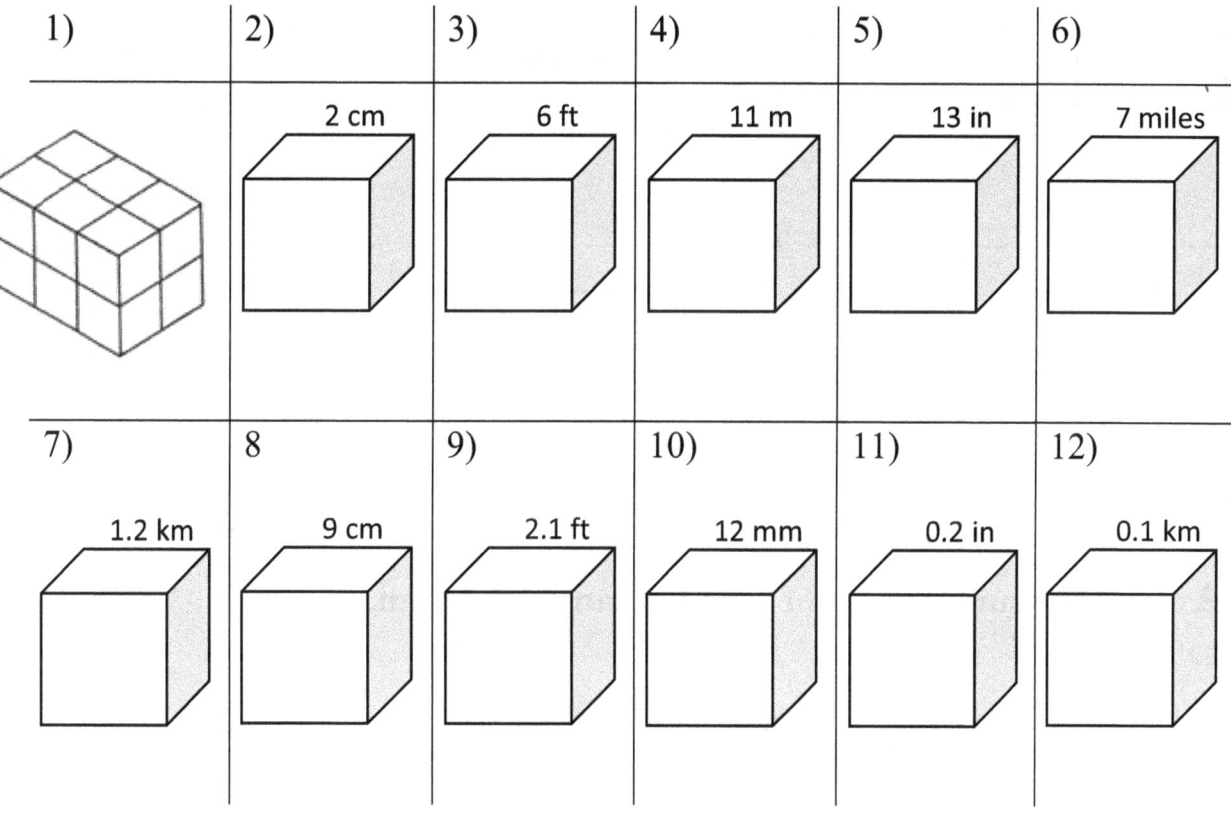

✏ Find the surface area of each cube.

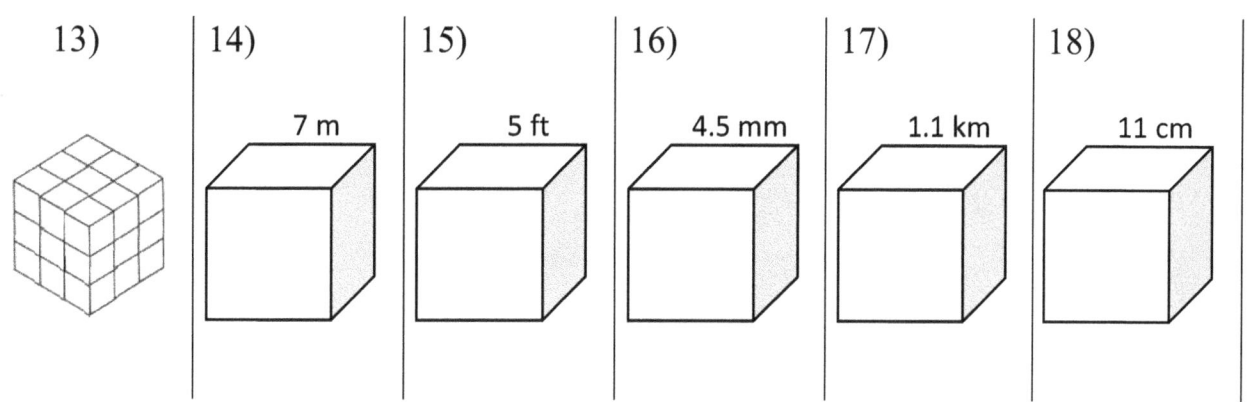

Rectangular Prism

✏ Find the volume of each Rectangular Prism.

WWW.MathNotion.Com

Common Core Subject Test Mathematics Grade 8

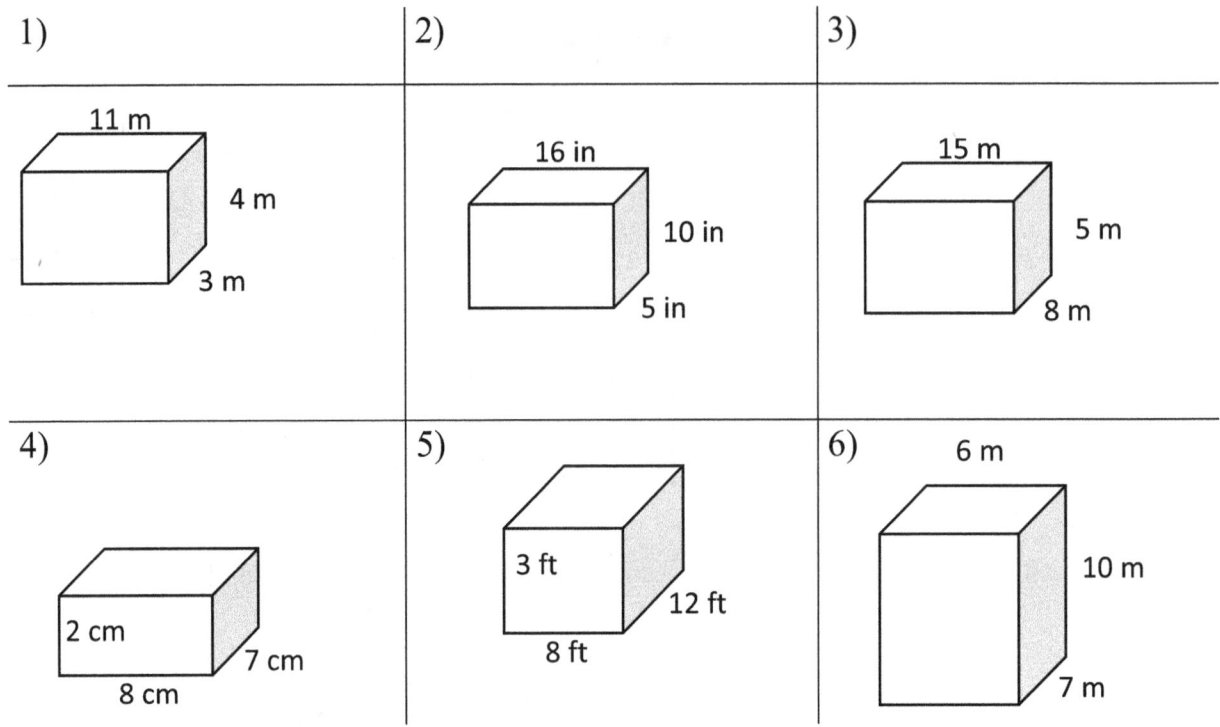

🖎 Find the surface area of each Rectangular Prism.

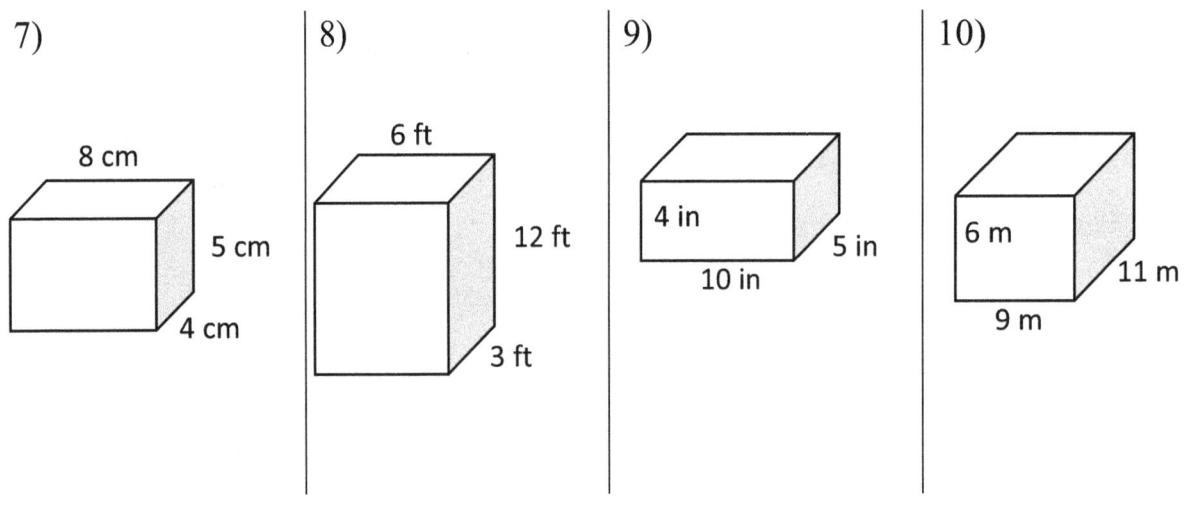

Common Core Subject Test Mathematics Grade 8

Cylinder

✏️ **Find the volume of each Cylinder. Round your answer to the nearest tenth.** ($\pi = 3.14$)

1)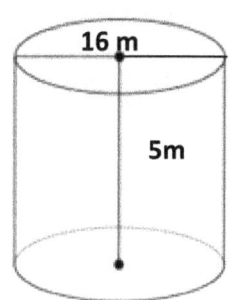

2)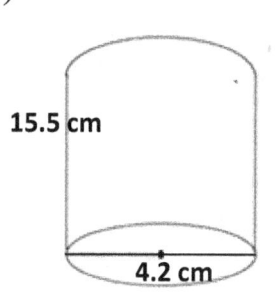

3)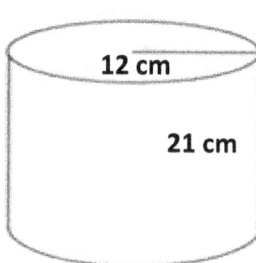

4)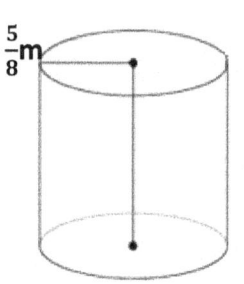

5)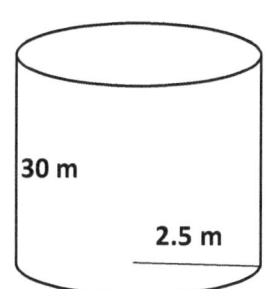

6)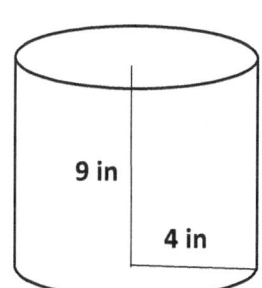

✏️ **Find the surface area of each Cylinder.** ($\pi = 3.14$)

7)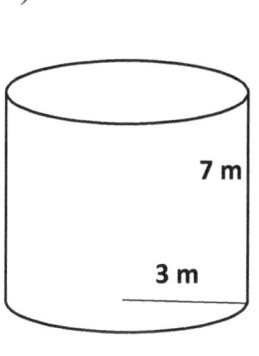

8)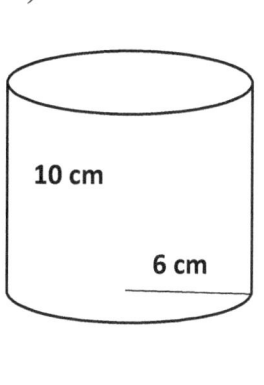

9)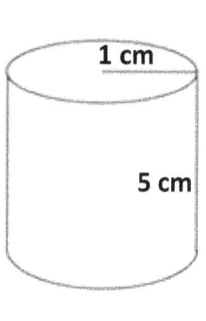

10)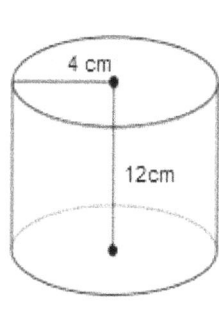

WWW.MathNotion.Com   139

Common Core Subject Test Mathematics Grade 8

Pyramids and Cone

✍ **Find the volume of each Pyramid and Cone.** ($\pi = 3.14$)

1)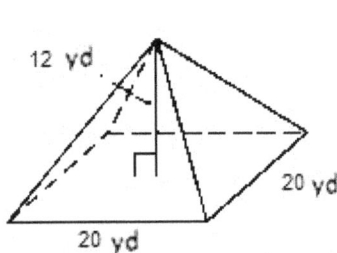

2)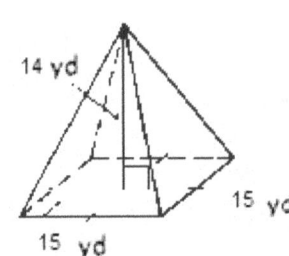

3)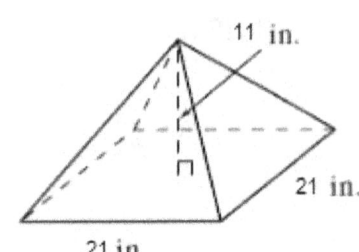

4)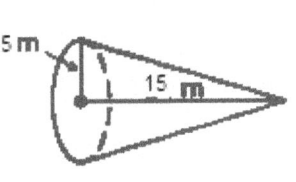

5)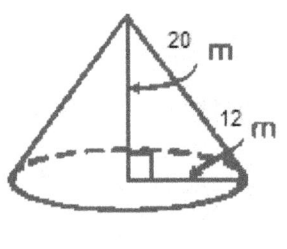

6)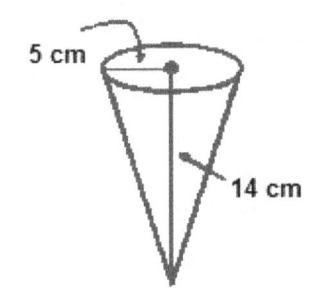

✍ **Find the surface area of each Pyramid and Cone.** ($\pi = 3.14$)

7)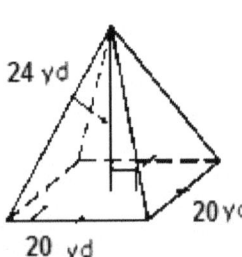

8)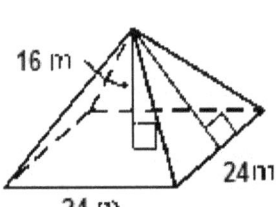

9)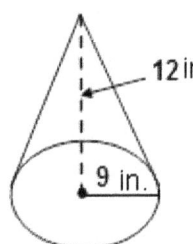

10)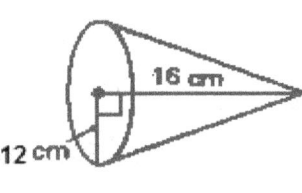

Common Core Subject Test Mathematics Grade 8

Answers of Worksheets

Angles

1) 16°   4) 34°   7) 90°   10) 75°
2) 96°   5) 70°   8) 75°   11) 70°
3) 59°   6) 52°   9) 33°

Pythagorean Relationship

1) No   5) Yes   9) 13   13) 15
2) Yes   6) No   10) 20   14) 30
3) No   7) Yes   11) 17   15) 36
4) Yes   8) Yes   12) 10   16) 12

Triangles

1) 60°   5) 45°   9) 54 square unites
2) 48°   6) 40°   10) 120 square unites
3) 55°   7) 45°   11) 90 square unites
4) 52°   8) 67°   12) 180 square unites

Polygons

1) 52 ft   5) 54 m   9) 30 m^2
2) 26 in   6) 25 cm   10) 300 in^2
3) 112 ft   7) 38 in   11) 160 km^2
4) 20 cm   8) 24 m   12) 49 in^2

Trapezoids

1) 50 cm^2 4) 60 cm^2 7) 36
2) 105 m^2 5) 80   8) 15
3) 39 ft^2 6) 24

Calculate

1) 13 cm   2) 11 ft   3) 12 m   4) 25 ft

Circles

1) 19.63 in^2 5) 379.94 cm^2 9) 12.56 m^2
2) 78.5 cm^2 6) 314 $miles^2$ 10) 113.04 cm^2
3) 254.34 ft^2 7) 132.67 in^2 11) 38.47 $miles^2$
4) 12.56 m^2 8) 7.07 ft^2 12) 50.24 ft^2

Common Core Subject Test Mathematics Grade 8

Circle No.	Radius	Diameter	Circumference	Area
1	1 in	2 in	6.28 in	3.14 in^2
2	5 m	10 m	31.4 m	78.5 m^2
3	3 ft	6 ft	18.84 ft	28.26 ft^2
4	7.5 miles	15 mi	47.1 mi	176.63 mi^2
5	5.5 km	11 km	34.54 km	94.99 km^2
6	7 cm	14 cm	43.96 cm	153.86 cm^2
7	6 ft	12 ft	37.68 feet	113.04 ft^2
8	10 m	20 m	62.8 m	314 m^2
9	9 in	18 in	56.52 in	254.34 in^2
10	4.5 ft	9 ft	28.26 ft	63.585 ft^2

Cubes

1) 12
2) 8 cm^3
3) 216 ft^3
4) 1,331 m^3
5) 2,197 in^3
6) 343 $miles^3$
7) 1.728 km^3
8) 729 cm^3
9) 9.261 ft^3
10) 1,728 mm^3
11) 0.008 in^3
12) 0.001 km^3
13) 27
14) 294 m^2
15) 150 ft^2
16) 121.5 mm^2
17) 7.26 km^2
18) 726 cm^2

Rectangular Prism

1) 132 m^3
2) 800 in^3
3) 600 m^3
4) 112 cm^3
5) 288 ft^3
6) 420 m^3
7) 184 cm^2
8) 252 ft^2
9) 220 in^2
10) 438 m^2

Cylinder

1) 1,004.8 m^3
2) 214.6 cm^3
3) 9,495.4 cm^3
4) 1.1 m^3
5) 588.8 m^3
6) 452.2 in^3
7) 188.4 m^2
8) 602.9 cm^2
9) 37.7 cm^2
10) 401.9 m^2

Pyramids and Cone

1) 1,600 yd^3
2) 1,050 yd^3
3) 1,617 in^3
4) 392.5 m^3
5) 3,014.4 m^3
6) 366.33 cm^3
7) 1,440 yd^2
8) 1,536 m^2
9) 678.24 in^2
10) 1,205.76 cm^2

Common Core Subject Test Mathematics Grade 8

Chapter 11 :
Statistics and Probability

Topics that you will practice in this chapter:

- ✓ Mean and Median
- ✓ Mode and Range
- ✓ Histograms
- ✓ Stem–and–Leaf Plot
- ✓ Pie Graph
- ✓ Probability Problems
- ✓ Factorials
- ✓ Combinations and Permutation

Mathematics is no more computation than typing is literature.
– John Allen Paulos

Common Core Subject Test Mathematics Grade 8

Mean and Median

🖎 Find Mean and Median of the Given Data.

1) 8, 7, 14, 4, 8

2) 14, 8, 25, 19, 16, 33, 11

3) 23, 18, 15, 12, 17

4) 34, 14, 10, 15, 6, 11

5) 10, 19, 6, 8, 32, 20, 17

6) 17, 26, 39, 69, 20, 6

7) 40, 38, 18, 11, 9, 2, 7, 32, 41

8) 24, 21, 31, 12, 33, 32, 22

9) 16, 14, 20, 41, 15, 20, 38, 4

10) 20, 20, 30, 18, 6, 28, 12, 46

11) 12, 7, 10, 11, 16, 22

12) 10, 29, 27, 12, 2, 15, 10, 3

🖎 Calculate.

13) In a javelin throw competition, five athletics score 56, 34, 62, 23 and 19 meters. What are their Mean and Median? _____

14) Eva went to shop and bought 8 apples, 14 peaches, 6 bananas, 4 pineapples and 12 melons. What are the Mean and Median of her purchase? _____

15) Bob has 17 black pen, 19 red pen, 14 green pens, 20 blue pens and 5 boxes of yellow pens. If the Mean and Median are 19 respectively, what is the number of yellow pens in each box? _____

Common Core Subject Test Mathematics Grade 8

Mode and Range

✎ **Find Mode and Rage of the Given Data.**

1) 4, 3, 7, 3, 3, 4
   Mode: _____   Range: _____

2) 18, 18, 24, 26, 18, 8, 14, 22
   Mode: _____   Range: _____

3) 8, 8, 8, 16, 19, 22, 20, 9, 13
   Mode: _____   Range: _____

4) 24, 24, 14, 28, 20, 18, 20, 24
   Mode: _____   Range: _____

5) 6, 21, 27, 24, 27, 27
   Mode: _____   Range: _____

6) 21, 8, 8, 7, 8, 12, 10, 22, 18, 13
   Mode: _____   Range: _____

7) 7, 4, 4, 6, 13, 13, 13, 0, 2, 2
   Mode: _____   Range: _____

8) 5, 8, 5, 14, 12, 14, 3, 5, 18
   Mode: _____   Range: _____

9) 7, 7, 7, 12, 7, 3, 8, 16, 3, 17
   Mode: _____   Range: _____

10) 15, 15, 19, 16, 4, 16, 10, 15
    Mode: _____   Range: _____

11) 6, 6, 5, 6, 42, 13, 19, 2
    Mode: _____   Range: _____

12) 8, 8, 9, 8, 9, 4, 34, 22
    Mode: _____   Range: _____

✎ **Calculate.**

13) A stationery sold 12 pencils, 56 red pens, 24 blue pens, 20 notebooks, 12 erasers, 21 rulers and 11 color pencils. What are the Mode and Range for the stationery sells?

    Mode: _____   Range: _____

14) In an English test, eight students score 10, 15, 15, 18 18, 16, 15 and 15. What are their Mode and Range? _____

15) What is the range of the first 6 even numbers greater than 8?
    _____

Common Core Subject Test Mathematics Grade 8

Times Series

🔖 Use the following Graph to complete the table.

Day	Distance (km)
1	
2	

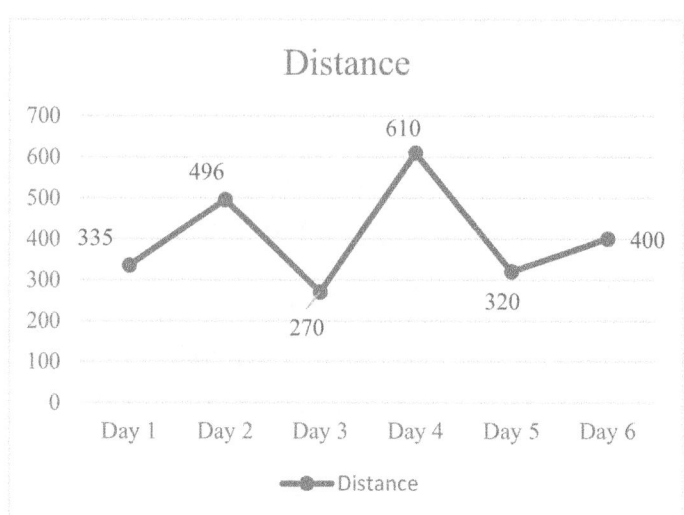

The following table shows the number of births in the US from 2007 to 2012 (in millions).

Year	Number of births (in millions)
2007	4.15
2008	3.70
2009	3.45
2010	3.20
2011	1.75
2012	2.98

Draw a Time Series for the table.

WWW.MathNotion.Com

Stem-and-Leaf Plot

✎ **Make stem ad leaf plots for the given data.**

1) 24, 26, 29, 20, 53, 27, 51, 55, 36, 21, 37, 30

   Stem | Leaf plot

2) 11, 59, 66, 14, 18, 19, 59, 65, 69, 61, 68, 65

   Stem | Leaf plot

3) 121, 55, 66, 54, 112, 128, 63, 125, 59, 123, 68, 119

   Stem | Leaf plot

4) 51, 32, 100, 56, 84, 36, 107, 56, 85, 39, 56, 106, 89

   Stem | Leaf plot

5) 33, 89, 19, 87, 81, 16, 11, 30, 86, 35, 17, 35, 13

   Stem | Leaf plot

6) 60, 92, 22, 25, 67, 93, 95, 62, 21, 64, 98, 29

   Stem | Leaf plot

WWW.MathNotion.Com

Pie Graph

The circle graph below shows all Robert's expenses for last month. Robert spent $140 on his hobbies last month.

Answer following questions based on the Pie graph.

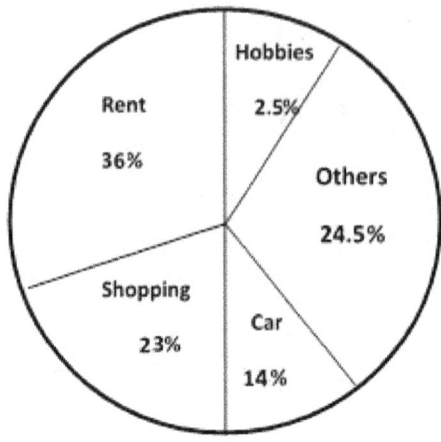

1) How much was Robert's total expenses last month? _____

2) How much did Robert spend on his car last month? _____

3) How much did Robert spend for shopping last month? _____

4) How much did Robert spend on his rent last month? _____

5) What fraction is Robert's expenses for his rent and car out of his total expenses last month? _____

Common Core Subject Test Mathematics Grade 8

Probability Problems

✎ **Calculate.**

1) A number is chosen at random from 1 to 10. Find the probability of selecting number 6 or smaller numbers. _____

2) Bag A contains 18 red marbles and 6 green marbles. Bag B contains 16 black marbles and 8 orange marbles. What is the probability of selecting a green marble at random from bag A? What is the probability of selecting a black marble at random from Bag B? _____

3) A number is chosen at random from 1 to 20. What is the probability of selecting multiples of 4? _____

4) A card is chosen from a well-shuffled deck of 52 cards. What is the probability that the card will be a queen? _____

5) A number is chosen at random from 1 to 15. What is the probability of selecting a multiple of 3 or 5? _____

A spinner numbered 1–8, is spun once. What is the probability of spinning …?

6) an Odd number? _____   7) a multiple of 2? _____

8) a multiple of 5? _____   9) number 10? _____

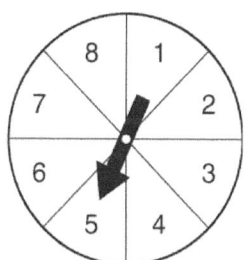

Common Core Subject Test Mathematics Grade 8

Factorials

✎ **Determine the value for each expression.**

1) $4! + 0! =$

2) $2! + 5! =$

3) $(2!)^2 =$

4) $5! - 3! =$

5) $6! - 3! + 10 =$

6) $3! \times 4 - 15 =$

7) $(2! + 3!)^2 =$

8) $(4! - 3!)^2 =$

9) $(3!\,0!)^2 - 10 =$

10) $\dfrac{10!}{8!} =$

11) $\dfrac{6!}{4!} =$

12) $\dfrac{6!}{5!} =$

13) $\dfrac{15!}{13!} =$

14) $\dfrac{n!}{(n-3)!} =$

15) $\dfrac{(n+2)!}{n!} =$

16) $\dfrac{(2+2!)^3}{2!} =$

17) $\dfrac{5(n+2)!}{(n+1)!} =$

18) $\dfrac{22!}{20!\,4!} =$

19) $\dfrac{13!}{11!\,3!} =$

20) $\dfrac{9 \times 210!}{3(7 \times 30)!} =$

21) $\dfrac{32!}{31!\,2!} =$

22) $\dfrac{11!\,12!}{10!\,13!} =$

23) $\dfrac{16!\,15!}{14!\,14!} =$

24) $\dfrac{(5 \times 3)!}{0!\,14!} =$

25) $\dfrac{4!(5n-2)!}{(5n)!} =$

26) $\dfrac{4n(4n+7)!}{(4n+8)!} =$

27) $\dfrac{(n-2)!(n+1)}{(n+2)!} =$

Common Core Subject Test Mathematics Grade 8

Combinations and Permutations

✍ **Calculate the value of each.**

1) 6! = ___

2) 2! × 5! = ___

3) 3 × 4! = ___

4) 5! + 3! = ___

5) 7! = ___

6) 4! = ___

7) 3! + 3! = ___

8) 7! − 5! = ___

✍ **Find the answer for each word problems.**

9) Susan is baking cookies. She uses sugar, butter, Vanilla, eggs and flour. How many different orders of ingredients can she try? _____

10) Albert is planning for his vacation. He wants to go to museum, watch a movie, go to the beach, play the game and play football. How many ways of ordering are there for him? _____

11) How many 4-digit numbers can be named using the digits 3, 4, 5, and 6 without repetition? _____

12) In how many ways can 5 boys be arranged in a straight line? _____

13) In how many ways can 6 athletes be arranged in a straight line? _____

14) A professor is going to arrange her 7 students in a straight line. In how many ways can she do this? _____

15) How many code symbols can be formed with the letters for the word GAMES? _____

16) In how many ways a team of 7 basketball players can choose a captain and co-captain? _____

WWW.MathNotion.Com

Common Core Subject Test Mathematics Grade 8

Answers of Worksheets

Mean and Median

1) Mean: 8.2, Median: 8
2) Mean: 18, Median: 16
3) Mean: 17, Median: 17
4) Mean: 15, Median: 12.5
5) Mean: 16, Median: 17
6) Mean: 29.5, Median: 23
7) Mean: 22, Median: 18
8) Mean: 25, Median: 24
9) Mean: 21, Median: 18
10) Mean: 22.5, Median: 20
11) Mean: 13, Median: 11.5
12) Mean: 13.5, Median: 11
13) Mean: 38.8, Median: 34
14) Mean: 8.8, Median: 8
15) 5

Mode and Range

1) Mode: 3, Range: 4
2) Mode: 18, Range: 18
3) Mode: 8, Range: 14
4) Mode: 24, Range: 14
5) Mode: 27, Range: 21
6) Mode: 8, Range: 15
7) Mode: 13, Range: 13
8) Mode: 5, Range: 15
9) Mode: 7, Range: 14
10) Mode: 15, Range: 15
11) Mode: 6, Range: 40
12) Mode: 8, Range: 30
13) Mode: 12, Range: 45
14) Mode: 15, Range: 8
15) 10

Time series

Day	Distance (km)
1	335
2	496
3	270
4	610
5	320
6	400

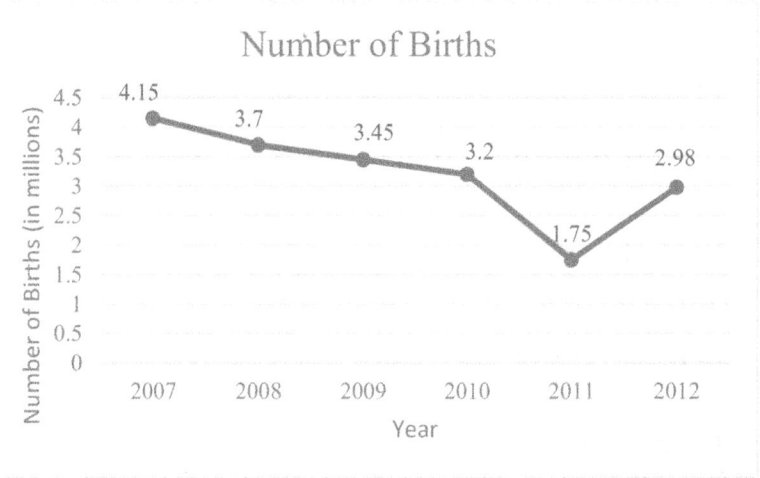

Stem–And–Leaf Plot

1)

Stem	leaf
2	0 1 4 6 7 9
3	0 6 7
5	1 3 5

2)

Stem	leaf
1	1 4 8 9
5	9 9
6	1 5 5 6 8 9

3)

Stem	leaf
5	4 5 9
6	3 6 8
11	2 9
12	1 3 5 8

Common Core Subject Test Mathematics Grade 8

4)

Stem	leaf
3	2 6 9
5	1 6 6 6
8	4 5 9
10	0 6 7

5)

Stem	leaf
1	1 3 6 7 9
3	0 3 5 5
8	1 6 7 9

6)

Stem	leaf
2	2 1 5 9
6	0 2 4 7
9	2 3 5 8

Pie Graph
1) $5,600
2) $784
3) $1,288
4) $2,016
5) $\frac{1}{2}$

Probability Problems
1) $\frac{3}{5}$
2) $\frac{1}{4}, \frac{2}{3}$
3) $\frac{1}{4}$
4) $\frac{1}{13}$
5) $\frac{7}{15}$
6) $\frac{1}{2}$
7) $\frac{1}{2}$
8) $\frac{1}{8}$
9) 0

Factorials
1) 25
2) 122
3) 4
4) 114
5) 724
6) 9
7) 64
8) 324
9) 26
10) 90
11) 30
12) 6
13) 210
14) $n(n-1)(n-2)$
15) $(n+1)(n+2)$
16) 32
17) $5(n+2)$
18) 19.25
19) 26
20) 3
21) 16
22) $\frac{11}{13}$
23) 3,600
24) 15
25) $\frac{24}{5n(5n-1)}$
26) $\frac{n}{(n+2)}$
27) $\frac{1}{n(n-1)(n+2)}$

Combinations and Permutations
1) 720
2) 240
3) 72
4) 126
5) 5,040
6) 24
7) 12
8) 4,920
9) 120
10) 120
11) 24
12) 120
13) 720
14) 5,040
15) 120
16) 42

Common Core Subject Test Mathematics Grade 8

Chapter 12 : Common Core Math Practice Tests

Time to Test

Time to refine your skill with a practice examination.

Take a REAL Common Core Mathematics test to simulate the test day experience. After you've finished, score your test using the answer key.

Before You Start

- You'll need a pencil, calculator, and a timer to take the test.
- It's okay to guess. You won't lose any points if you're wrong.
- After you've finished the test, review the answer key to see where you went wrong.

Graphing calculators are permitted for Common Core Tests Grade 8

Good Luck!

Common Core Subject Test Mathematics Grade 8

Common Core Subject Test Mathematics Grade 8

Common Core GRADE 8 MAHEMATICS REFRENCE MATERIALS

Linear Equations

Slope-intercept form	$y = mx + b$
Slope of a line	$m = \frac{y_2 - y_1}{x_2 - x_1}$

Circumference

Circle	$C = 2\pi r$	or	$C = \pi d$

Area

Triangle	$A = \frac{1}{2} bh$
Rectangle or Parallelogram	$A = bh$
Trapezoid	$A = \frac{1}{2} h(b_1 + b_2)$
Circle	$A = \pi r^2$

Surface Area

	Lateral	Total
Prism	$S = ph$	$S = ph + 2B$
Cylinder	$S = 2\pi rh$	$S = 2\pi rh + 2\pi r^2$

Volume

Prism or cylinder	$V = Bh$
Pyramid or Cone	$V = \frac{1}{3} Bh$
Sphere	$V = \frac{4}{3} \pi r^3$

Additional Information

Pythagorean theorem	$a^2 + b^2 = c^2$
Simple interest	$I = prt$
Compound interest	$I = p(1 + r)^t$

Common Core Subject Test Mathematics Grade 8

Common Core Subject Test Mathematics Grade 8

Common Core Practice Test 1

Mathematics

GRADE 8

Administered Month Year

Common Core Subject Test Mathematics Grade 8

1) The area of a circle is 36π. Which of the following can be the circumference of the circle?

   A. 12π C. 36π

   B. 24π D. 18π

2) You can buy 17 cans of green beans at a supermarket for $4.30. How much does it cost to buy 51 cans of green beans?

   A. $16.6   C. $12.9

   B. $162.2   D. $129.9

3) Which of the following is the solution of the following inequality?

$$-8x + 20.5 > -11x + 6.5 + 6.5x$$

   A. $x < 4$ C. $x < -4$

   B. $x > 4$ D. $x > -4$

4) What is the perimeter of a square that has an area of 249.64 feet?

   Write your answer in the box below.

   ☐

5) A tree 18 feet tall casts a shadow 44 feet long. Jack is 9 feet tall. How long is Jack's shadow?

   A. 18 ft   C. 22 ft

   B. 28 ft   D. 42 ft

WWW.MathNotion.Com

Common Core Subject Test Mathematics Grade 8

6) The price of a laptop is decreased by 45% to $610. What is its original price?

   A. 1109.09

   B. 119.09

   C. 209.90

   D. 1199.9

7) A container holds 6.3 gallons of water when it is $\frac{7}{34}$ full. How many gallons of water does the container hold when it's full?

   A. 30.6

   B. 43

   C. 18.65

   D. 16.3

8) If $(5^a)^b = 625$, then what is the value of $a \times b$?

   A. $5a$

   B. $3b$

   C. 4

   D. 3

9) A bag contains 22 balls: six green, three black, three blue, nine red and one white. If 15 balls are removed from the bag at random, what is the probability that a white ball has been removed?

   A. $\frac{3}{22}$

   B. $\frac{15}{22}$

   C. $\frac{1}{22}$

   D. $\frac{1}{15}$

10) What is the x-intercept of the line with equation $-8x + 9y = 72$?

    A. -8

    B. -9

    C. $+8$

    D. $-\frac{8}{9}$

Common Core Subject Test Mathematics Grade 8

11) Which of the following expressions is equivalent to

$6x^4y(7x + 3y)$?

A. $18yx^5 - 42x^4y$

B. $42x^3 - 18y^3$

C. $42yx^5 + 18x^4y^2$

D. $42x^4y^2 - 18yx^5$

12) A soccer team played 170 games and won 20 percent of them. How many games did the team win?

A. 120

B. 34

C. 64

D. 60

13) The equation of a line is given as: $y = -3x + 2$. Which of the following points does not lie on the line?

A. $(2, -4)$

B. $(3, -7)$

C. $(-2, 8)$

D. $(4, -5)$

14) The perimeter of the trapezoid below is 34 cm. What is its area?

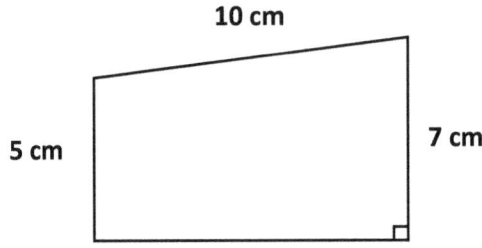

Write your answer in the box below.

Common Core Subject Test Mathematics Grade 8

15) Which graph does not represent y as a function of x?

A.   B.

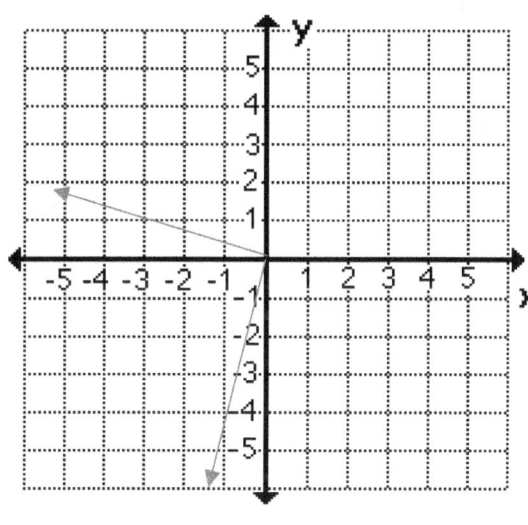

C.   D.

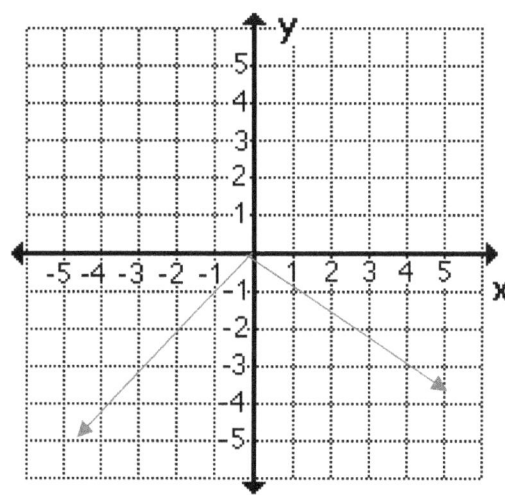 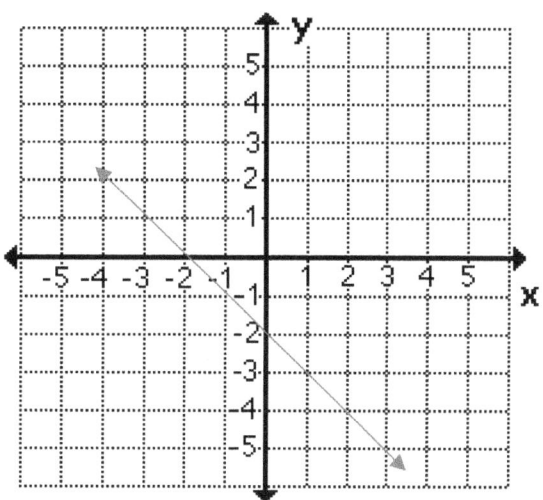

16) Which of the following is equivalent to $-31 < -6x + 5 < 5$?

A. $0 < x < 6$ C. $-6 < x < 0$

B. $1 < x < 6$ D. $-6 < x < 6$

WWW.MathNotion.Com

Common Core Subject Test Mathematics Grade 8

17) A bank is offering 1.95% simple interest on a savings account. If you deposit $25,000, how much interest will you earn in six years?

A. $2,925

C. $2,450

B. $1,950

D. $4,725

18) Joe scored 33 out of 43 marks in Algebra, 42 out of 51 marks in science and 18 out of 29 marks in mathematics. In which subject his percentage of marks is best?

A. Mathematics

C. Algebra

B. Science

D. Mathematics and Science

19) What is the volume of the following triangular prism?

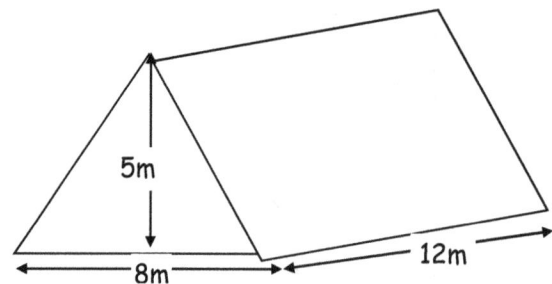

Write your answer in the box below.

20) The marked price of a computer is E Euro. Its price decreased by 20% in March and later increased by 8 % in April. What is the final price of the computer in E Euro?

A. 0.864 E

C. 0.84 E

B. 0.096E

D. 0.968 E

Common Core Subject Test Mathematics Grade 8

21) Find the slope–intercept form of the graph $3x - 9y = 18$

    A. $y = 3x + 2$ C. $y = \frac{1}{3}x - 2$

    B. $y = -\frac{1}{3}x + 2$ D. $y = -\frac{1}{3}x - 2$

22) Triangle ABC is graphed on a coordinate grid with vertices at A (1, 4), B (–2, 8) and C (2, 9). Triangle ABC is reflected over x axes to create triangle A' B' C'. Which order pair represents the coordinate of C'?

    A. (9, 2)   C. (−2, −9)

    B. (2, −9)   D. (−2, 9)

23) Which of the following point is the solution of the system of equations?

$$\begin{cases} 7x - 5y = 5 \\ 3x + 2y = -2 \end{cases}$$

    A. $(-2, 0)$ C. $(-1, 1)$

    B. $(0, -1)$ D. $(2, -1)$

24) A shirt costing $400 is discounted 34%. After a month, the shirt is discounted another 16%. Which of the following expressions can be used to find the selling price of the shirt?

    A. (400) (0.66) (0.16)

    B. (400) (0.84) (0.16)

    C. (400) (0.66) (0.84)

    D. (400) (0.34) (0.16)

Common Core Subject Test Mathematics Grade 8

25) What is the distance between the points (2, 1) and (− 6, − 5)?

    A. 10

    B. 15

    C. 20

    D. 25

Questions 26 and 27 are based on the following data

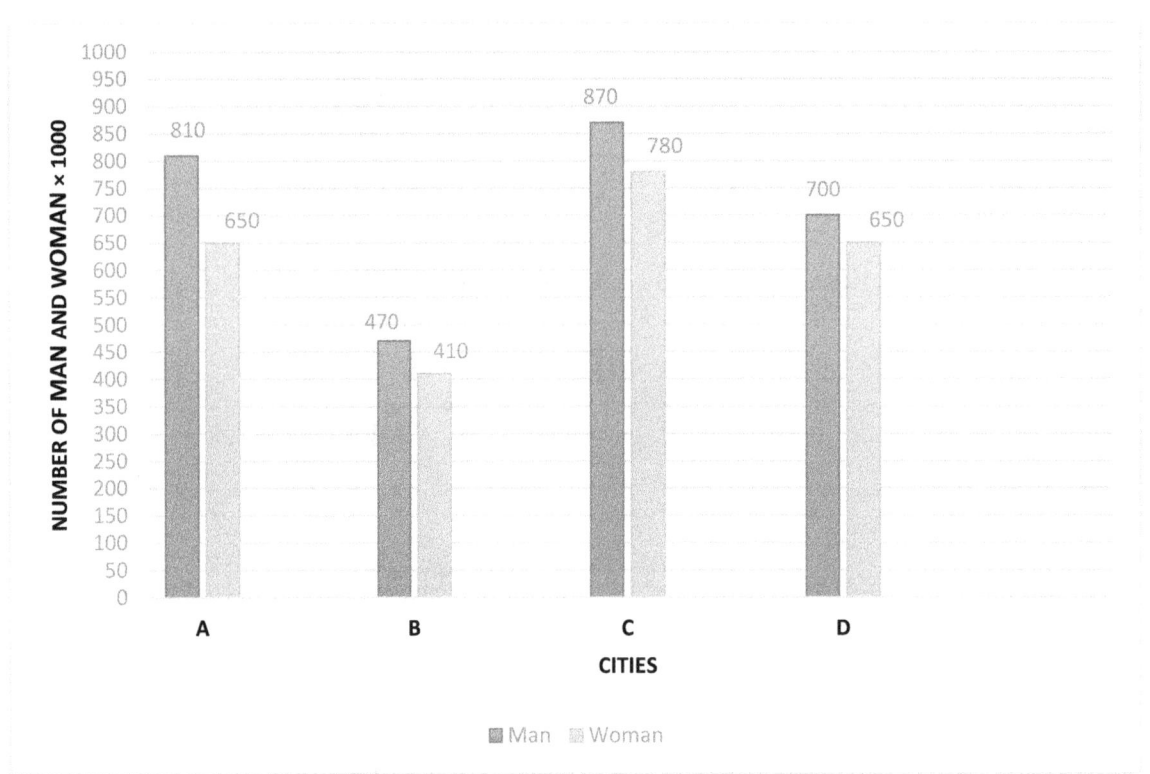

26) What's the ratio of percentage of men in city C to percentage of women in city B?

    A. 0.099   C. 1.015

    B. 0.985   D. 1.05

Common Core Subject Test Mathematics Grade 8

27) What's the maximum ratio of woman to man in the four cities?

   A. 0.93

   B. 0.80

   C. 0.78

   D. 0.75

28) Line m passes through the point $(2, -8)$. Which of the following **CANNOT** be the equation of line m?

   A. $y = -8$

   B. $y = -5x + 2$

   C. $y = -4 + x$

   D. $y = -6x + 4$

29) The sum of four numbers is 89. If another number is added to these four numbers, the average of the five numbers is 34. What is the fifth number?

   A. 101

   B. 93

   C. 87

   D. 81

30) David owed $14,749. After making 43 payments of $257 each, how much did he have left to pay?

   A. $3,869

   B. $4,230

   C. $3,698

   D. $4,689

31) Which of the following lines is parallel to the graph of $y = 7x - 4$?

   A. $7x - 3y = 4$

   B. $4x - 7y = 0$

   C. $7x - y = -13$

   D. $4x + y = 11$

Common Core Subject Test Mathematics Grade 8

32) The average of five consecutive numbers is 31. What is the smallest number?

    A. 29   C. 27

    B. 25   D. 32

33) The price of a laptop is decreased by 36% to $288. What is its original price?

    A. 450   C. 460

    B. 420.5   D. 250

34) The average weight of 13 girls in a class is 42 kg and the average weight of 17 boys in the same class is 45 kg. What is the average weight of all the 30 students in that class?

    A. 42.77 Kg   C. 47.33 Kg

    B. 43.46Kg   D. 43.7 Kg

35) Liam's average (arithmetic mean) on five mathematics tests is 61. What should Liam's score be on the next test to have an overall of 59 for all the tests?

    A. 49   C. 58

    B. 60   D. 50

36) Which of the following equations has a graph that is a straight line?

    A. $y = 4x^2 - 11$ C. $x - y = 11$

    B. $7y^2 - 6x^2 = 5$ D. $2x + xy = 11$

Common Core Subject Test Mathematics Grade 8

37) An angle is equal to one fourth of its supplement. What is the measure of that angle?

   A. 45

   B. 15

   C. 36

   D. 20

38) Six seventh of 35 is equal to $\frac{3}{5}$ of what number?

   A. 38

   B. 52

   C. 54

   D. 50

39) A card is drawn at random from a standard 52–card deck, what is the probability that the card is of Hearts? (The deck includes 13 of each suit clubs, diamonds, hearts, and spades)

   A. $\frac{1}{4}$

   B. $\frac{1}{13}$

   C. $\frac{3}{52}$

   D. $\frac{4}{13}$

40) The score of Zoe was one fourth of Emma and the score of Harper was triple that of Emma. If the score of Harper was 48, what is the score of Zoe?

   A. 12

   B. 4

   C. 24

   D. 8

Practice Test 1

This is the End of this Section.

Common Core Subject Test Mathematics Grade 8

Common Core Subject Test Mathematics Grade 8

Common Core Practice Test 2

Mathematics

GRADE 8

Administered Month Year

Common Core Subject Test Mathematics Grade 8

1) Right triangle ABC has two legs of lengths 12 cm (AB) and 16 cm (AC). What is the length of the third side (BC)?

   A. 14 cm

   B. 22 cm

   C. 18 cm

   D. 20 cm

2) When a number is subtracted from 64 and the difference is divided by that number, the result is 7. What is the value of the number?

   A. 8

   B. 4

   C. 10

   D. 18

3) A bank is offering 3.6% simple interest on a savings account. If you deposit $13,000, how much interest will you earn in two years?

   A. $196

   B. $1,336

   C. $369

   D. $936

4) In a party, 14 soft drinks are required for every 8 guests. If there are 152 guests, how many soft drinks is required?

   A. 136

   B. 266

   C. 160

   D. 340

5) A chemical solution contains 18% alcohol. If there is 7.2 ml of alcohol, what is the volume of the solution?

   A. 45.5ml

   B. 55 ml

   C. 40 ml

   D. 65 ml

Common Core Subject Test Mathematics Grade 8

6) What is the area of the shaded region?

   A. 65

   B. 130

   C. 120

   D. 55

   8 ft · 11 ft · 5 ft · 15 ft

7) A rope weighs 340 grams per meter of length. What is the weight in kilograms of 18.6 meters of this rope? (1 kilograms = 1000 grams)

   A. 6,324

   B. 0.6324

   C. 63.24

   D. 6.324

8) The ratio of boys to girls in a school is 3: 7. If there are 330 students in a school, how many boys are in the school.

   Write your answer in the box below.

9) In two successive years, the population of a town is increased by 9% and 16%. What percent of the population is increased after two years?

   A. 26.12%

   B. 26.44%

   C. 126.44%

   D. 64.4 %

Common Core Subject Test Mathematics Grade 8

10) Which graph shows a non–proportional linear relationship between *x* and *y*?

A.

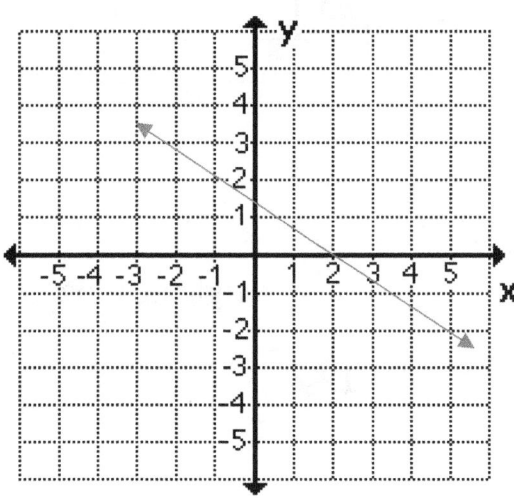

B.

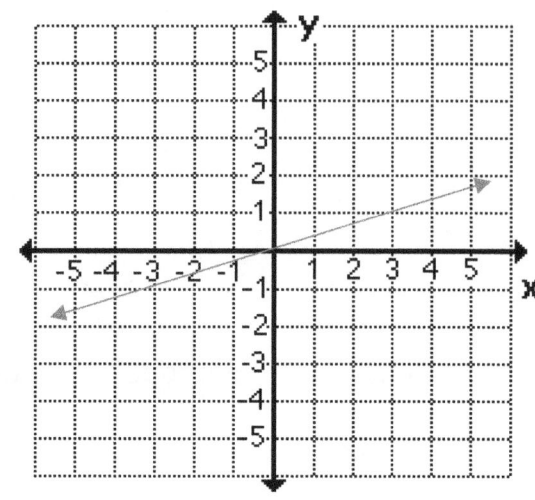

C.

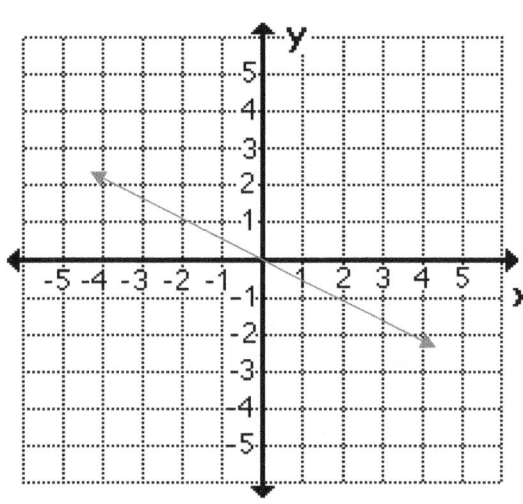

D.

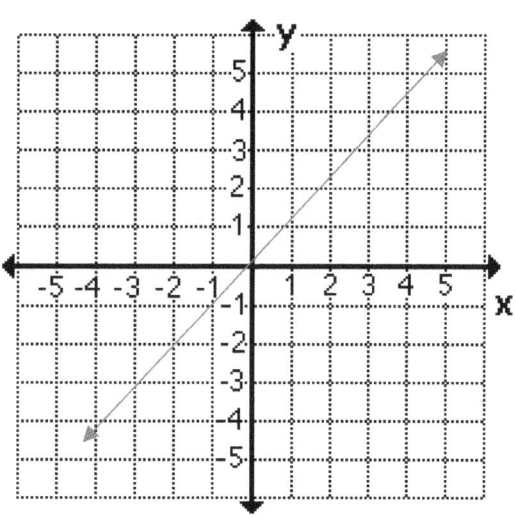

11) Three years ago, Amy was five times as old as Mike was. If Mike is 7 years old now, how old is Amy?

A. 20

B. 23

C. 16

D. 26

Common Core Subject Test Mathematics Grade 8

12) What is the value of $|-52+19| - |7(-4)|$?

   A. 5

   B. -5

   C. -12

   D. $+12$

13) What is the solution of the following system of equations?

$$\begin{cases} \dfrac{x}{8} + \dfrac{y}{4} = 3 \\ \dfrac{-7y}{8} - x = -6 \end{cases}$$

   A. $x = -8, y = 16$

   B. $x = -4, y = 18$

   C. $x = 6, y = 18$

   D. $x = -8, y = 18$

14) If a gas tank can hold 56 gallons, how many gallons does it contain when it is $\dfrac{5}{8}$ full?

   A. 45

   B. 35

   C. 15

   D. 25

15) What is the length of BC in the following figure if AB = 4, DF = 5 and BD = 36?

   A. 12

   B. 16

   C. 8

   D. 32

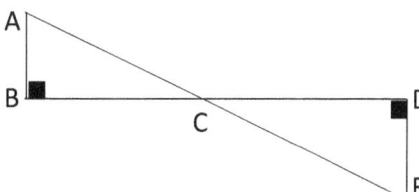

Common Core Subject Test Mathematics Grade 8

16) In the xy-plane, the point $(5, 3)$ and $(6, 7)$ are on the line A. Which of the following equations of lines is parallel to line A?

   A. $4y - 2x = 3$

   B. $y = \frac{3x}{2} + \frac{7}{2}$

   C. $3y = 5x + 9$

   D. $3y - 12x = 9$

17) In the rectangle below if $y > 5$ cm and the area of rectangle is 36 cm² and the perimeter of the rectangle is 26 cm, what is the value of x and y respectively?

   A. 4, 9

   B. 3, 12

   C. 11, 4

   D. 6, 6

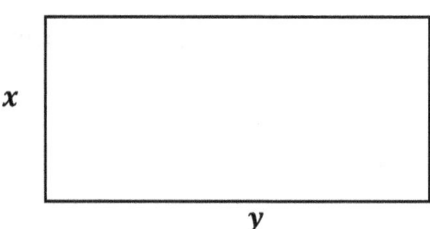

18) A football team had $33,000 to spend on supplies. The team spent $24,000 on new balls. New sport shoes cost $114 each. Which of the following inequalities represent the number of new shoes the team can purchase?

   A. $114x + 24,000 \leq 33,000$

   B. $114x + 18,000 \geq 33,000$

   C. $24,000 - 114x \geq 33,000$

   D. $24,000 - 114x \leq 33,000$

Common Core Subject Test Mathematics Grade 8

19) A $60 shirt now selling for $48 is discounted by what percent?

   A. 15 %

   B. 10 %

   C. 20 %

   D. 25 %

20) How much interest is earned on a principal of $14,000 invested at an interest rate of 1.8% for five years?

   A. $1,260

   B. $2,360

   C. $1,240

   D. $12,420

21) The price of a car was $80,000 in 2014, $52,000 in 2015 and $33,800 in 2016. What is the rate of depreciation of the price of car per year?

   A. 35 %

   B. 30 %

   C. 25 %

   D. 15 %

22) The Jackson Library is ordering some bookshelves. If x is the number of bookshelves the library wants to order, which each cost $52 and there is a one-time delivery charge of $720, which of the following represents the total cost, in dollar, per bookshelf?

   A. $52x + 720$

   B. $52 + 720x$

   C. $\dfrac{52x+720}{52}$

   D. $\dfrac{52x+720}{x}$

Common Core Subject Test Mathematics Grade 8

23) The following table represents the value of x and function $f(x)$. Which of the following could be the equation of the function $f(x)$?

A. $f(x) = 3x^2 + 5$

B. $f(x) = 3x^2 - 4x + 7$

C. $f(x) = \sqrt{3x + 5}$

D. $f(x) = 5\sqrt{x} + 8$

x	$f(x)$
0	8
4	18
9	23
16	28

24) The circle graph below shows all Mr. Wilson's expenses for last month. If he spent $640 on his car, how much did he spend for his rent?

A. $1,360

B. $1,600

C. $4,000

D. 4,200

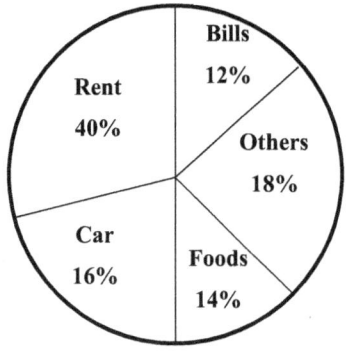

Mr. Wilson's Monthly Expenses

25) The sum of five different negative integers is −80. If the smallest of these integers is −18, what is the largest possible value of one of the other five integers?

A. −14

B. −13

C. −8

D. −16

Common Core Subject Test Mathematics Grade 8

26) In the following figure, point M lies on the line A, what is the value of y if $x = 10$?

A. 15

B. 12

C. 13

D. 8

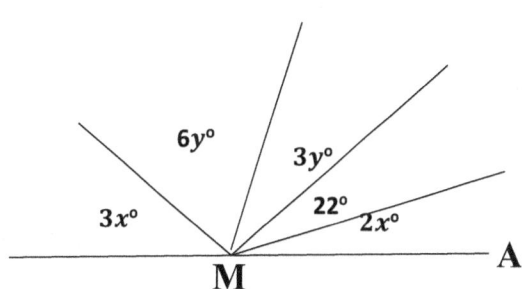

27) What is the smallest integer whose square root is greater than 9?

A. 144   C. 121

B. 83   D. 88

28) What is the area of the shaded region if the diameter of the bigger circle is 14 inches and the diameter of the smaller circle is 8 inches?

A. 22 π

B. 34 π

C. 33π

D. 32 π

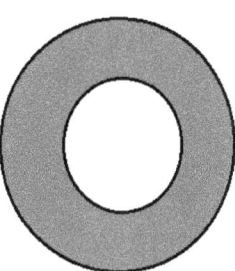

29) What is the sum of $\sqrt{2x+9}$ and $\sqrt{x}-2$ when $\sqrt{x} = -6$?

A. 4   C. 12

B. 5   D. 13

WWW.MathNotion.Com

Common Core Subject Test Mathematics Grade 8

30) What is the area of an isosceles right triangle that has one leg that measures 18 cm?

Write your answer in the box below.

31) If x is directly proportional to the square of y, and $y = 4$ when $x = 80$, then if $x = 180$ $y = ?$

A. 6

B. 7

C. 9

D. 36

32) A swimming pool holds 1,600 cubic feet of water. The swimming pool is 40 feet long and 8 feet wide. How deep is the swimming pool?

Write your answer in the box below.

33) What is the value of x in the following figure?

A. 57

B. 72

C. 59

D. 63

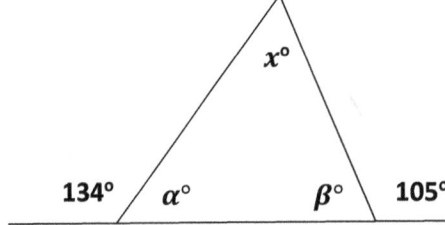

Common Core Subject Test Mathematics Grade 8

34) Jack earns $450 for his first 30 hours of work in a week and is then paid 1.2 times his regular hourly rate for any additional hours. This week, Jack needs $666 to pay his rent, bills and other expenses. How many hours must he work to make enough money in this week?

A. 15

B. 30

C. 27

D. 32

Questions 35, 36 and 37 are based on the following data

Types of air pollutions in 10 cities of a country

Type of Pollution	Number of Cities
A	
B	
C	
D	
E	
	1 2 3 4 5 6 7 8 9 10

35) If a is the mean (average) of the number of cities in each pollution type category, b is the mode, and c is the median of the number of cities in each pollution type category, then which of the following must be true?

A. $a = c < b$

B. $b = a = c$

C. $c < a$

D. $a < b < c$

Common Core Subject Test Mathematics Grade 8

36) How many cities should be added to type of pollutions B until the ratio of cities in type of pollution B to cities in type of pollution D will be 1.25?

   A. 10

   B. 6

   C. 8

   D. 14

37) What percent of cities are in the type of pollution A, C, and E respectively?

   A. 70%, 30%, 50%

   B. 1.70%, 1.30%, 1.50%

   C. 1.50%, 1.70%, 1.30%

   D. 30%, 50%, 1.20%

38) In the following right triangle, if the sides AB and AC become one-third longer, what will be the ratio of the perimeter of the triangle to its area?

   A. $\frac{1}{6}$

   B. $\frac{1}{3}$

   C. 3

   D. 6

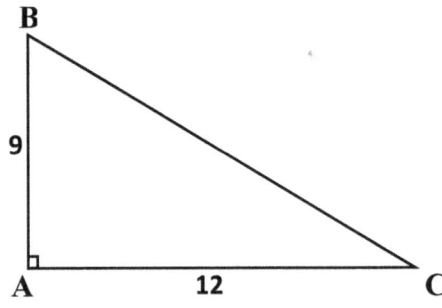

39) The capacity of a red box is 25% bigger than the capacity of a blue box. If the red box can hold 75 equal sized books, how many of the same books can the blue box hold?

   A. 25

   B. 40

   C. 60

   D. 50

Common Core Subject Test Mathematics Grade 8

40) A taxi driver earns $12 per hour work. If he works 8 hours a day, and he uses 5-liters Petrol in 2.5 hours with price $2.4 for 1-liter. How much money does he earn in one day?

   A. $96

   B. $62.5

   C. $38.4

   D. $57.6

Practice Test 2

This is the End of this Section.

Common Core Subject Test Mathematics Grade 8

Chapter 13 : Answers and Explanations

Common Core Practice Tests

Answer Key

※ Now, it's time to review your results to see where you went wrong and what areas you need to improve!

Common Core - Mathematics

Practice Test - 1

1	A	**16**	A	**31**	C
2	C	**17**	A	**32**	A
3	A	**18**	B	**33**	A
4	63.2	**19**	240	**34**	D
5	C	**20**	A	**35**	A
6	A	**21**	C	**36**	C
7	A	**22**	B	**37**	C
8	C	**23**	B	**38**	D
9	C	**24**	C	**39**	A
10	B	**25**	A	**40**	B
11	C	**26**	C		
12	B	**27**	B		
13	D	**28**	C		
14	72	**29**	D		
15	A	**30**	C		

Practice Test - 2

1	D	**16**	D	**31**	A
2	A	**17**	A	**32**	5
3	D	**18**	A	**33**	C
4	B	**19**	C	**34**	C
5	C	**20**	A	**35**	C
6	A	**21**	A	**36**	B
7	D	**22**	D	**37**	A
8	99	**23**	D	**38**	C
9	B	**24**	B	**39**	C
10	A	**25**	A	**40**	D
11	B	**26**	C		
12	A	**27**	B		
13	A	**28**	C		
14	C	**29**	D		
15	B	**30**	162		

Common Core Subject Test Mathematics Grade 8

Common Core Subject Test Mathematics Grade 8

Practice Test 1
Answers and Explanations

1) Answer: A

Use the formula for area of circles.

Area = $\pi r^2 \Rightarrow 36\pi = \pi r^2 \Rightarrow 36 = r^2 \Rightarrow r = 6$

Radius of the circle is 6. Now, use the circumference formula:

Circumference = $2\pi r = 2\pi (6) = 12\pi$

2) Answer: C

Let x be the number of cans. Write the proportion and solve for x.

$\frac{17 \text{ cans}}{\$4.30} = \frac{51 \text{ cans}}{x} \Rightarrow x = \frac{4.30 \times 51}{17} \Rightarrow x = \$12.9$

3) Answer: A

$-8x + 20.5 > -11x + 6.5 + 6.5x \rightarrow$ Combine like terms:

$-8x + 20.5 > -4.5x + 6.5$ Subtract $8x$ from both sides: $20.5 > 3.5x + 6.5$

Add -6.5 both sides of the inequality.

$14 > 3.5x$, Divide both sides by 3.5. $\Rightarrow \frac{14}{3.5} > x \rightarrow x < 4$

4) Answer: 63.2 feet.

Area of a square: $S = a^2 \Rightarrow 249.64 = a^2 \Rightarrow a = 15.8$

Perimeter of a square: $P = 4a \Rightarrow P = 4 \times 15.8 \Rightarrow P = 63.2$

5) Answer: C

Write the proportion and solve for the missing number.

$\frac{18}{44} = \frac{9}{x} \rightarrow 18x = 9 \times 44 = 396 \rightarrow 18x = 396 \rightarrow x = \frac{396}{18} = 22$

6) Answer: A

Let x be the original price. If the price of a laptop is decreased by 45% to $610, then:

$55\% \text{ of } x = 610 \Rightarrow 0.55x = 610 \Rightarrow x = 610 \div 0.55 = 1109.09$

7) Answer: A

let x be the number of gallons of water the container holds when it is full.

Then; $\frac{7}{34}x = 6.3 \rightarrow x = \frac{34 \times 6.3}{7} = 30.6$

WWW.MathNotion.Com

Common Core Subject Test Mathematics Grade 8

8) Answer: C

$(5^a)^b = 625 \to 5^{ab} = 625 \to 625 = 5^4 \to 5^{ab} = 5^4 \to ab = 4$

9) Answer: C

If 15 balls are removed from the bag at random, there will be one ball in the bag. The probability of choosing a white ball is 1 out of 22. Therefore, the probability of not choosing a white ball is 15 out of 22 and the probability of having not a white ball after removing 15 balls is the same.

10) Answer: B

The value of y in the x-intercept of a line is zero. Then:

$y = 0 \to -8x + 9(0) = 72 \to -8x = 72 \to x = \frac{-72}{8} = -9$

then, x-intercept of the line is -9

11) Answer: C

Use distributive property:

$6x^4y(7x + 3y) = 6x^4y(7x) + 6x^4y(3y) = 42yx^5 + 18x^4y^2$

12) Answer: B

$170 \times \frac{20}{100} = 34$.

13) Answer: D

$y = -3x + 2$

$(2, -4) \Rightarrow -4 = -3(2) + 2 \Rightarrow -4 = -4$

$(3, -7) \Rightarrow -7 = -3(3) + 2 \Rightarrow -7 = -7$

$(-2, 8) \Rightarrow 8 = -3(-2) + 2 \Rightarrow 8 = 8$

$(4, -5) \Rightarrow -5 = -3(4) + 2 \Rightarrow -10 \neq -5$

14) Answer: 72.

The perimeter of the trapezoid is 34 cm.

Therefore, the missing side (height) is = 34 − 10 − 5 − 7 = 12

Area of a trapezoid: A = $\frac{1}{2}$ h ($b_1 + b_2$) = $\frac{1}{2}$ (12) (5 + 7) = 72

15) Answer: A

A graph represents y as a function of x if $x_1 = x_2 \to y_1 = y_2$

Common Core Subject Test Mathematics Grade 8

In choice A, for each x, we have two different values for y.

16) Answer: A

$-31 < -6x + 5 < 5 \rightarrow$ Subtract 5 to all sides.

$-31 - 5 < -6x + 5 - 5 < 5 - 5 \rightarrow -36 < -6x < 0 \rightarrow$ Divide all sides by -6.

(Remember that when you divide all sides of an inequality by a negative number, the inequality sing will be swapped. $<$ becomes $>$)

$\frac{-36}{-6} < \frac{-6x}{-6} < \frac{0}{-6} \Rightarrow 6 > x > 0$, or $0 < x < 6$

17) Answer: A

Use simple interest formula: $I = prt$ (I = interest, p = principal, r = rate, t = time)

$I = (25,000)(0.0195)(6) = 2,925$

18) Answer: B

Compare each mark:

In Algebra Joe scored 33 out of 43 in Algebra. It means Joe scored 76.74% of the total mark. $\frac{33}{43} = \frac{x}{100} \Rightarrow x = 76.74\%$

Joe scored 42 out of 51 in science. It means Joe scored 82.35% of the total mark. $\frac{42}{51} = \frac{x}{100} \Rightarrow x = 82.35\%$

Joe scored 18 out of 29 in mathematic that it means 62.06% of total mark. $\frac{18}{29} = \frac{x}{100} \Rightarrow x = 62.06\%$

Therefore, his score in Science is higher than his other scores.

19) Answer: 240 m³.

Use the volume of the triangular prism formula. $V = \frac{1}{2}$ (length) (base) (high)

$V = \frac{1}{2} \times 12 \times 8 \times 5 \Rightarrow V = 240$ m³

20) Answer: A

To find the discount, multiply the number by (100% – rate of discount).

Therefore, for the first discount we get: $(100\% - 20\%)(E) = (0.80)E$

For increase of 8 %:

$(0.80)E \times (100\% + 8\%) = (0.80)(1.08) = 0.864E$

Common Core Subject Test Mathematics Grade 8

21) Answer: C

$3x - 9y = 18 \Rightarrow -9y = -3x + 18 \Rightarrow y = \frac{-3}{-9}x + \frac{18}{-9} \Rightarrow y = \frac{1}{3}x - 2$

$y = \frac{1}{3}x - 2$

22) Answer: B

When a point is reflected over x axes, the (y) coordinate of that point changes to $(-y)$ while its x coordinate remains the same.

B $(2, 9) \to$ B' $(2, -9)$

23) Answer: B

Solving Systems of Equations by Elimination

$\begin{array}{l} 7x - 5y = 5 \\ 3x + 2y = -2 \end{array}$ Multiply the first equation by -3, and second equation by 7, then add two equations.

$\begin{array}{l} -3(7x - 5y = 5) \\ 7(3x + 2y = -2) \end{array} \Rightarrow \begin{array}{l} -21x + 15y = -15 \\ 21x + 14y = -14 \end{array} \Rightarrow 29y = -29 \Rightarrow y = -1.$

$3x + 2y = -2, 3x + 2(-1) = -2$, then: $x = 0, (0, -1)$

24) Answer: C

To find the discount, multiply the number by (100% – rate of discount).

Therefore, for the first discount we get: $(400)(100\% - 34\%) = (400)(0.66)$

For the next 16 % discount: $(400)(0.66)(0.84)$.

25) Answer: A

Use distance formula:

$C = \sqrt{(x_A - x_B)^2 + (y - y_B)^2} \Rightarrow C = \sqrt{(2-(-6))^2 + (1-(-5))^2}$

$C = \sqrt{(8)^2 + (6)^2} \Rightarrow C = \sqrt{64 + 36} \Rightarrow C = \sqrt{100} = 10$

26) Answer: C

Percentage of women in city C $= \frac{780}{1,650} \times 100 = 47.27\%$

Percentage of men in city B $= \frac{410}{880} \times 100 = 46.59\%$

Percentage of women in city C to percentage of men in city B: $\frac{47.27}{46.59} = 1.015$

Common Core Subject Test Mathematics Grade 8

27) Answer: B

Ratio of women to men in city A: $\frac{650}{810} = 0.80$

Ratio of women to men in city B: $\frac{410}{470} = 0.87$

Ratio of women to men in city C: $\frac{780}{870} = 0.897$

Ratio of women to men in city D: $\frac{650}{700} = 0.93$

28) Answer: C

Solve for each equation: $(2, -8)$

$y = -8 \Rightarrow -8 = -8$

$y = -5x + 2 \Rightarrow -8 = -5(2) + 2 \Rightarrow -8 = -8$

$y = -4 + x \Rightarrow -8 = -4 + 2 \Rightarrow -2 \neq -8$

$y = -6x + 4 \Rightarrow -8 = -6(2) + 4 \Rightarrow -8 = -8$

29) Answer: D

$a + b + c + d = 89 \Rightarrow \frac{a+b+c+d+e}{5} = 34 \Rightarrow a + b + c + d + e = 170$

$\Rightarrow 89 + e = 170 \Rightarrow e = 170 - 89 = 81$

30) Answer: C

$43 \times \$257 = \$11,051$ Payable amount is: $\$14,749 - 11,051 = 3,698$

31) Answer: C

If two lines are parallel with each other, then the slope of the two lines is the same.

Then in line $y = 7x - 4$, the slope is equal to 7

And in the line $7x - y = -13 \Rightarrow y = 7x + 13$, the slope equal to 7

32) Answer: A

Let x be the smallest number. Then, these are the numbers:

$x, x + 1, x + 2, x + 3, x + 4$

average $= \frac{\text{sum of terms}}{\text{number of terms}} \Rightarrow 31 = \frac{x+(x+1)+(x+2)+(x+3)+(x+4)}{5} \Rightarrow 31 = \frac{5x+10}{5}$

$\Rightarrow 155 = 5x + 10 \Rightarrow c \Rightarrow x = 29$

33) Answer: A

Let x be the original price.

Common Core Subject Test Mathematics Grade 8

If the price of a laptop is decreased by 36% to $288, then:

$64\% \text{ of } x = 288 \Rightarrow 0.64\ x = 288 \Rightarrow x = 288 \div 0.64 = 450$

34) Answer: D

The sum of the weight of all girls is: $13 \times 42 = 546$ kg

The sum of the weight of all boys is: $17 \times 45 = 765$ kg

The sum of the weight of all students is: $546 + 765 = 1,311$ kg

$\text{average} = \dfrac{\text{sum of terms}}{\text{number of terms}}$; $\text{average} = \dfrac{1,311}{30} = 43.7$

35) Answer: A

$\dfrac{a+b+c+d+e}{5} = 61 \Rightarrow a+b+c+d+e = 305$

$\dfrac{a+b+c+d+e+f}{6} = 59 \Rightarrow a+b+c+d+e+f = 354$

$305 + f = 354 \Rightarrow f = 354 - 305 = 49$

36) Answer: C

$x - y = 11$ has a graph that is a straight line. All other options are not equations of straight lines.

37) Answer: C

The sum of supplement angles is 180. Let x be that angle. Therefore,

$x + 4x = 180$; $5x = 180$, divide both sides by 5: $x = 36$

38) Answer: D

Let x be the number. Write the equation and solve for x.

$\dfrac{6}{7} \times 35 = \dfrac{3}{5} x \Rightarrow \dfrac{6 \times 35}{7} = \dfrac{3x}{5}$, use cross multiplication to solve for x.

$30 \times 35 = 3x \times 7 \Rightarrow 1,050 = 21x \Rightarrow x = 50$

39) Answer: A

The probability of choosing a Hearts is $\dfrac{13}{52} = \dfrac{1}{4}$

40) Answer: B

If the score of Harper was 48, therefore the score of Emma is 16. Since the score of Zoe was one fourth of Emma, therefore, the score of Zoe is 4.

Common Core Subject Test Mathematics Grade 8

Practice Test 2

Answers and Explanations

1) Answer: D

Use Pythagorean Theorem: $a^2 + b^2 = c^2$

$12^2 + 16^2 = C^2 \Rightarrow 144 + 256 = C^2 \Rightarrow 400 = c^2 \Rightarrow c = 20$

2) Answer: A

Let x be the number. Write the equation and solve for x.

$\frac{(64-x)}{x} = 7$ (cross multiply)

$(64 - x) = 7x$, then add x both sides. $64 = 8x$, now divide both sides by 8. $\Rightarrow x = 8$.

3) Answer: D

Use simple interest formula: $I = prt$;

($I = interest. p = principal. r = rate. t = time$)

$I = (13,000)(0.036)(2) = 936$

4) Answer: B

Let x be the number of soft drinks for 152 guests. Write the proportion and solve for x.

$\frac{14 \text{ soft drinks}}{8 \text{ guests}} = \frac{x}{152 \text{ guests}} \Rightarrow x = \frac{152 \times 14}{8} \Rightarrow x = 266$

5) Answer: C

18% of the volume of the solution is alcohol. Let x be the volume of the solution. Then:

18% of x = 7.2 ml

$0.18 \, x = 7.2 \Rightarrow \frac{18x}{100} = \frac{72}{10}$ cross multiply; $180x = 7,200 \Rightarrow$ (devide by 180) $x = 40$

6) Answer: A

Use the area of rectangle formula (S = a × b).

To find area of the shaded region subtract smaller rectangle from bigger rectangle.

$S_1 - S_2 = (8 \, ft \times 15 ft) - (5 ft \times 11 ft) \Rightarrow S_1 - S_2 = 65 ft$.

7) Answer: D

The weight of 18.6 meters of this rope is: $18.6 \times 340 g = 6,324 g$

1 kg = 1,000 g, therefore, 6,324 $g \div 1000 = 6.324$ kg

Common Core Subject Test Mathematics Grade 8

8) Answer: 99

The ratio of boy to girls is 3:7. Therefore, there are 3 boys out of 10 students. To find the answer, first divide the total number of students by 10, then multiply the result by 3.

$330 \div 10 = 33 \Rightarrow 33 \times 3 = 99$

9) Answer: B

the population is increased by 9% and 16%. 9% increase changes the population to 109% of original population.

For the second increase, multiply the result by 116%.

$(1.09) \times (1.16) = (1.2644) = 126.44\%$

26.44 percent of the population is increased after two years.

10) Answer: A

A linear equation is a relationship between two variables, x and y, that can be put in the form $y = mx + b$.

A non-proportional linear relationship takes on the form $y = mx + b$, where $b \neq 0$ and its graph is a line that does not cross through the origin.

11) Answer: B

Three years ago, Amy was five times as old as Mike. Mike is 7 years now. Therefore, 3 years ago Mike was 4 years.

Three years ago, Amy was: $A = 4 \times 5 = 20$

Now Amy is 23 years old: $20 + 3 = 23$

12) Answer: A

$|-52 + 19| - |7(-4)| = |-33| - |-28| = 33 - 28 = 5$

13) Answer: A

$\begin{cases} \frac{x}{8} + \frac{y}{4} = 3 \\ \frac{-7y}{8} - x = -6 \end{cases}$ → Multiply the top equation by 8. Then,

$\begin{cases} x + 2y = 24 \\ \frac{-7y}{8} - x = -6 \end{cases}$ → Add two equations.

$\frac{9}{8}y = 18 \rightarrow y = 16$, plug in the value of y into the first equation → $x = -8$

Common Core Subject Test Mathematics Grade 8

14) Answer: C

$\frac{5}{8} \times 56 = \frac{280}{8} = 35$

15) Answer: B

Two triangles $\triangle ABC$ and $\triangle CD$ are similar. Then:

$\frac{AB}{DF} = \frac{BC}{CD} \rightarrow \frac{4}{5} = \frac{x}{36-x} \rightarrow 144 - 4x = 5x \rightarrow 9x = 144 \rightarrow x = 16$

16) Answer: D

The slop of line A is: $m = \frac{y_2 - y_1}{x_2 - x_1} = \frac{7-3}{6-5} = 4$

Parallel lines have the same slope and only choice D ($3y - 12x = 9 \Rightarrow y = 4x + 3$) has slope of 4.

17) Answer: A

The perimeter of the rectangle is: $2x + 2y = 26$

$\rightarrow x + y = 13 \rightarrow x = 13 - y$

The area of the rectangle is: $x \times y = 36 \rightarrow (13 - y)(y) = 36$

$\rightarrow y^2 - 13y + 36 = 0$

Solve the quadratic equation by factoring method.

$(y - 4)(y - 9) = 0 \rightarrow y = 4$ (Unacceptable, because y must be greater than 5) or $y = 9$; If $y = 9 \rightarrow x = 13 - y \rightarrow x = 13 - 9 \rightarrow x = 4$

18) Answer: A

Let x be the number of new shoes the team can purchase. Therefore, the team can purchase $114\ x$.

The team had $33,000 and spent $24,000. Now the team can spend on new shoes $9,000 at most.

Now, write the inequality: $114x + 24,000 \leq 33,000$

19) Answer: C

Use the formula for Percent of Change

$\frac{New\ Value - Old\ Value}{Old\ Value} \times 100\ \% \Rightarrow \frac{48 - 60}{60} \times 100\ \% = -20\ \%$

Common Core Subject Test Mathematics Grade 8

(negative sign here means that the new price is less than old price).

20) Answer: A

Use simple interest formula: $I = prt$ (I = interest, p = principal, r = rate, t = time)

$I = (14{,}000)(0.018)(5) = 1{,}260$

21) Answer: A

Use this formula: Percent of Change $= \dfrac{New\ Value - Old\ Value}{Old\ Value} \times 100\ \%$

$\dfrac{52{,}000 - 80{,}000}{80{,}000} \times 100\ \% = -35\ \%$ and $\dfrac{33{,}800 - 52{,}000}{52{,}000} \times 100\% = -35\ \%$

22) Answer: D

The amount of money for x bookshelf is: $52x$

Then, the total cost of all bookshelves is equal to: $52x + 720$

The total cost, in dollar, per bookshelf is: $\dfrac{Total\ cost}{number\ of\ items} = \dfrac{52x + 720}{x}$

23) Answer: D

A. $f(x) = 3x^2 + 5$ if $x = 4 \to f(4) = 3(4)^2 + 5 = 53 \neq 18$

B. $f(x) = 3x^2 - 4x + 7$ if $x = 4 \to f(4) = 3(4)^2 - 4 \times 4 + 7 = 39 \neq 18$

C. $f(x) = \sqrt{3x + 5}$ if $x = 4 \to f(4) = \sqrt{3(4) + 5} = \sqrt{17} \neq 18$

D. $f(x) = 5\sqrt{x} + 8$ if $x = 4 \to f(4) = 5\sqrt{4} + 8 = 18 = 18$

24) Answer: B

Let x be all expenses, then $\dfrac{16}{100}x = \$640 \to x = \dfrac{100 \times \$640}{16} = \$4{,}000$

He spent for his rent: $\dfrac{40}{100} \times \$4{,}000 = \$1{,}600$

25) Answer: A

The smallest number is -18. To find the largest possible value of one of the other five integers, we need to choose the smallest possible integers for four of them. Let x be the largest number. Then:

$-80 = (-18) + (-17) + (-16) + (-15) + x \to -80 = -66 + x$

$\to x = -80 + 66 = -14$

Common Core Subject Test Mathematics Grade 8

26) Answer: C

The angles on a straight line add up to 180 degrees. Then:

$2x + 22 + 3y + 6y + 3x = 180$, Then

$5x + 9y = 180 - 22 \rightarrow 5(10) + 9y = 158 \rightarrow 9y = 158 - 50 = 108 \rightarrow y = 12$

27) Answer: B

Square root of 144 is $\sqrt{144} = 12 > \sqrt{81} = 9$

Square root of 83 is $\sqrt{83} = \sqrt{81+2} > \sqrt{81} = 9$

Square root of 121 is $\sqrt{121} = 11 > 9$

Square root of 88 is $\sqrt{88} = \sqrt{81+7} > 9$

Since $\sqrt{83} < \sqrt{88}$, then the answer is B.

28) Answer: C

To find the area of the shaded region subtract smaller circle from bigger circle.

$S_{bigger} - S_{smaller} = \pi (r_{bigger})^2 - \pi (r_{smaller})^2$

$\Rightarrow S_{bigger} - S_{smaller} = \pi (7)^2 - \pi (4)^2 \Rightarrow 49\pi - 16\pi = 33\pi$

29) Answer: D

$\sqrt{x} = -6 \rightarrow x = 36$

then; $\sqrt{x} - 2 = \sqrt{36} - 2 = 6 - 2 = 4$ and $\sqrt{2x+9} = \sqrt{72+9} = \sqrt{81} = 9$ Then:

$(\sqrt{2x+9}) + (\sqrt{x} - 2) = 9 + 4 = 13$

30) Answer: 162.

$a = 18 \Rightarrow$ area of the triangle is

$= \frac{1}{2}(18 \times 18) = \frac{324}{2} = 162 \; cm^2$

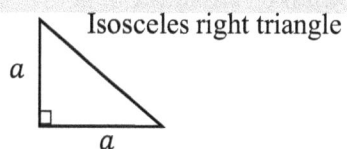

Isosceles right triangle

31) Answer: A

x is directly proportional to the square of y. Then: $x = cy^2$

$80 = c(4)^2 \rightarrow 80 = 16c \rightarrow c = \frac{80}{16} = 5$

The relationship between x and y is: $x = 5y^2$, $x = 180$

$180 = 5y^2 \rightarrow y^2 = \frac{180}{5} = 36 \rightarrow y = 6$

Common Core Subject Test Mathematics Grade 8

32) Answer: 5.

Use formula of rectangle prism volume.

V = (length) (width) (height) ⇒ 1,600 = (40) (8) (height) ⇒ height = 1,600 ÷ 320 = 5

33) Answer: C

$\alpha = 180° - 134° = 46°$

$\beta = 180° - 105° = 75°$

$x + \alpha + \beta = 180° \rightarrow x = 180° - 46° - 75° = 59°$

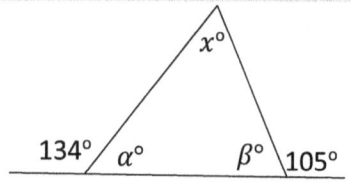

34) Answer: C

The amount of money that jack earns for one hour: $\frac{\$450}{30} = \$15$

Number of additional hours that he works to make enough money is: $\frac{\$666-\$450}{1.2\times\$15} = 12$

Number of total hours is: $15 + 12 = 27$

35) Answer: C

Let's find the mean (average), mode and median of the number of cities for each type of pollution.

Number of cities for each type of pollution: 7, 4, 3, 8, 5

average (mean) = $\frac{sum\ of\ terms}{number\ of\ terms} = \frac{7+4+3+8+5}{5} = \frac{27}{5} = 5.4$

Median is the number in the middle. To find median, first list numbers in order from smallest to largest: 3, 4, 5, 7, 8

Median of the data is 5.

Mode is the number which appears most often in a set of numbers. Therefore, there is no mode in the set of numbers.

Median < Mean, then, $c < a$

36) Answer: B

Let the number of cities should be added to type of pollutions B be x. Then:

$\frac{x+4}{8} = 1.25 \rightarrow x + 4 = 8 \times 1.25 \rightarrow x + 4 = 10 \rightarrow x = 6$

37) Answer: A

Percent of cities in the type of pollution A: $\frac{7}{10} \times 100 = 70\%$

Common Core Subject Test Mathematics Grade 8

Percent of cities in the type of pollution C: $\frac{3}{10} \times 100 = 30\%$

Percent of cities in the type of pollution E: $\frac{5}{10} \times 100 = 50\%$

38) Answer: C

$AB = 9$, And $AC = 12$

$BC = \sqrt{9^2 + 12^2} = \sqrt{81 + 144} = \sqrt{225} = 15$

Perimeter $= 9 + 12 + 15 = 36$; Area $= \frac{9 \times 12}{2} = 54$

In this case, the ratio of the perimeter of the triangle to its area is: $\frac{36}{54} = \frac{2}{3}$

If the sides AB and AC become half longer, then: $AB = 3$, And $AC = 4$

$BC = \sqrt{3^2 + 4^2} = \sqrt{9 + 16} = \sqrt{25} = 5$

Perimeter $= 3 + 4 + 5 = 12$; Area $= \frac{3 \times 4}{2} = 3 \times 2 = 6$

In this case the ratio of the perimeter of the triangle to its area is: $\frac{12}{6} = 2$

39) Answer: C

The capacity of a red box is 25% bigger than the capacity of a blue box and it can hold 75 books. Therefore, we want to find a number that 25% bigger than that number is 75. Let x be that number. Then:

$1.25 \times x = 75$, Divide both sides of the equation by 1.25. Then: $x = \frac{75}{1.25} = 60$

40) Answer: D

$\$12 \times 8 = \$96$

Petrol use: $5 \div 2.5 = 2$. $8 \times 2 = 16$ liters

Petrol cost: $16 \times \$2.4 = \$38.4$

Money earned: $\$96 - \$38.4 = \$57.6$

"End"

www.ingramcontent.com/pod-product-compliance
Lightning Source LLC
Chambersburg PA
CBHW080438110426
42743CB00016B/3201